VLSI Signal Processing Systems

VLSI Signal Processing Systems

by

Earl E. Swartzlander, Jr.
TRW

Kluwer Academic Publishers
Boston/Dordrecht/Lancaster

Distributors for North America:
Kluwer Academic Publishers
190 Old Derby Street
Hingham, Massachusetts 02043, USA

Distributors for the UK and Ireland:
Kluwer Academic Publishers
MTP Press Limited
Falcon House, Queen Square
Lancaster LA1 1RN, UNITED KINGDOM

Distributors for all other countries:
Kluwer Academic Publishers Group
Distribution Centre
Post Office Box 322
3300 AH Dordrecht, THE NETHERLANDS

Consulting Editor: Jonathan Allen

Library of Congress Cataloging-in-Publication Data
Swartzlander, Earl E.

  VLSI signal processing systems.

  Bibliography: p.
  Includes index.
  1. Signal processing—Digital techniques.
2. Integrated circuits—Very large scale integration.
I. Title.
TK5102.5.S957   1986   621.38'0414   85-20147
ISBN 0-89838-207-6

Copyright © 1986 by Kluwer Academic Publishers

All rights reserved. No part of this publication may be reproduced, stored in a retrieval system, or transmitted in any form or by any means, mechanical, photocopying, recording, or otherwise, without written permission of the publisher, Kluwer Academic Publishers, 190 Old Derby Street, Hingham, Massachusetts 02043.

Printed in the United States of America

CONTENTS

Preface		vii
1.	The VLSI Context	1
2.	VLSI Design	11
3.	VLSI Architecture	39
4.	Signal Processing	67
5.	Digital Filter Case Study	89
6.	Frequency Domain Filter Case Study	117
7.	Custom VLSI Case Study	141
8.	Signal Processing Networks	159
Author Index		181
Subject Index		185

PREFACE

During the last decade the technical community has witnessed a revolution in two superficially unrelated areas: VLSI development and digital signal processing. Close examination shows that they are not unrelated and that there is a high degree of synergy between them. Specifically, advances in VLSI have made it possible to successfully implement significant digital signal processing systems. The demonstrated utility of such digital signal processors has created a demand for even higher VLSI performance levels.

The purpose of this book is to provide architectural guidance to designers of signal processing systems and modules and the VLSI circuits needed to implement them. A related purpose is to illustrate successful application of the architectural concepts to real signal processing systems. Toward this goal, the book examines the algorithms, structures, and technology appropriate for development of special purpose systems. This strong applications orientation manifests itself in the inclusion of numerous examples and case studies. Such material is intended to provide motivation, indicate practical issues that arise in the development of real systems, and establish an intuitive feeling for the current state-of-the-art.

The first three chapters address VLSI from an internal and external architectural perspective. The next four chapters examine signal processing applications and implementation of example systems with VLSI. Finally, Chapter 8 provides an introduction to networking. Generalized networks are not now widely used for signal processing systems, but as the networking field matures and as more signal processing systems become distributed (both physically and logically), the two fields will merge in much the same way that VLSI and signal processing have merged over the past decade.

This book is an outgrowth of a series of seminars presented in the United States and in Europe in the Spring of 1983. In fact, the seminars were based on work performed (and published) over the preceding decade. Subsequent to the seminars, many subsets of the material have been presented in tutorials, workshops, and technical briefings. Throughout this process many constructive comments have been made by the audiences that have greatly improved the presentation of the material.

Based on feedback from the various audiences, the book should be useful to engineers and managers working in VLSI and digital signal processing, and should be an effective supplement for graduate classes in signal processing. The book is oriented toward system and logic design. No attempt has been made to cover either VLSI processing or specific signal processing algorithm development.

A minor notational note is appropriate: throughout the book, log is used to indicate the base 2 logarithm while ln indicates the natural logarithm.

I would like to acknowledge my sincere gratitude to Lauren Hall who transcribed the original lectures and typed innumerable copies of the manuscript with a consistent cheerful attitude. Tracy Blyth and the support staff in TRW publications have provided significant assistance during the final manuscript preparation process. During the last decade, TRW has provided continued support and encouragement for my various activities in VLSI and signal processing. Such support is truly gratifying. Most of all I thank my wife, Joan, who radiates joy, happiness, and enthusiasm.

VLSI Signal Processing Systems

CHAPTER 1.
THE VLSI CONTEXT

In the last 25 years a revolution in electronics has occurred. Component technology has evolved from discrete transistors to Very Large Scale Integrated (VLSI) circuits containing tens of thousands of transistors on a single chip. This chapter describes the changing technology environment in Section 1.1. The succeeding sections briefly describe the VLSI design process and the goals of the Very High Speed Integrated Circuit (VHSIC) program.

1.1 Technology Evolution

In 1964 Gordon Moore noted that the number of components per circuit for the most advanced integrated circuits had doubled every year since 1959 and predicted that the trend would continue [1-1]. More recently, he has modified "Moore's Law" showing an annual doubling from 1959 to 1975 and predicting a biannual doubling subsequent to 1975 [1-2] as charted in Figure 1-1.

It is convenient to categorize technology into four levels of integration as shown in Table 1-1. The complexity spans a four order of magnitude range. In the early to mid 1960s, Small Scale Integration (SSI) devices were being developed with 1 to 30 gates on a chip. At the Medium Scale Integration (MSI) level the number

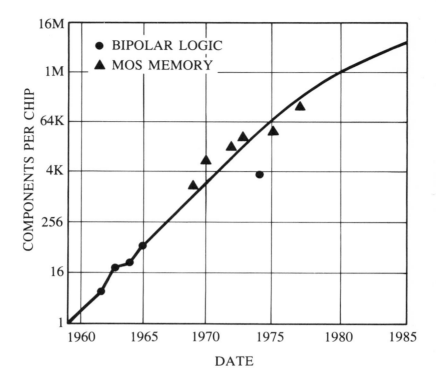

Figure 1-1. Moore's Law (After [1-2])

Integration Level	Introduced	Gates/Chips
Small Scale (SSI)	1964	1 to 30
Medium Scale (MSI)	1968	10 to 300
Large Scale (LSI)	1972	100 to 3K
Very Large Scale (VLSI)	1976	1K to 30K

Table 1-1. Historical Evolution

of gates on a chip increased by an order of magnitude to about 100 gates per chip. The Large Scale Integration (LSI) level began in the early 1970s with another order of magnitude increase in gate count. Early in the "LSI era" designers of electronic systems began to develop custom circuits tailored to the requirements of specific systems to reduce complexity and improve performance. Complex systems, consisting of many circuit boards of SSI and MSI devices could be reduced to a few custom chips thus simplifying the processor and reducing the production cost, power consumption, weight, etc. At the current VLSI level where circuits consist of tens of thousands of gates, the high nonrecurring cost of custom circuit development requires that the development effort is amortized across many projects. In the current environment a chip that is developed for a particular system will be designed with interfaces facilitating its use in other systems. Thus the chip development cost can be shared amongst the many systems that use it.

Table 1-2 shows typical circuit functions developed with the various levels of technology. The issues that have been the major concern to users at each functional level are identified. Early SSI device functions were gates and flip flops, i.e., simple traditional logic building blocks. These are the basic logic and memory functions that are taught in logic design classes. The companies that developed SSI circuits based their designs on the functions implemented with vacuum tube and transistor logic modules in early generations of digital computers. The largest user-issue then was critical mass; specifically, does a particular chip family have all of the functions needed to build a wide variety of systems? Will the logic designer get into a situation where the processor can be built except for one critical component that doesn't exist or is very difficult to implement? Some early families suffered from the lack of critical logic functions.

Integration	Function	User Issues
SSI	Gates, Flip-Flops	Critical Mass
MSI	Counters, 4 Bit Adders	Breadth
LSI	Memory, Microprocessors	Programming
VLSI	Signal Processors, Multipliers	System "Glue"

Table 1-2. Functional Evolution

The lack of critical components was effectively solved in the development of MSI, which was created by adding functions to the successful SSI families. The new functions were multiple adders, 4-bit wide binary, decimal, and duo-decimal counters, multiplexers, etc. These devices extended beyond previous computer modules into a higher level of complexity and were designed to complement the previous SSI circuits. The user's concern was which family had the widest breadth and would be most efficient in developing a variety of different systems. When MSI became available, most logic was standardized to Transistor-Transistor Logic (TTL) interfaces, so that users could mix components from several companies.

Within LSI, the most significant developments were memory and the microprocessor. Semiconductor Random Access Memory (RAM) was the first circuit to achieve LSI (and subsequently VLSI) levels of complexity. Dynamic RAM is a nearly ideal vehicle for technology demonstration since RAM circuits are developed by replicating a few cell types in two dimensional patterns. As a result the circuit design effort required to develop a new generation RAM is much less than the effort to develop a new logic circuit.

The microprocessor was a new system component unlike anything previously used. Users had to program the microprocessors to realize specific applications; a double edged sword. The microprocessor user achieves high flexibility because it is possible to reprogram the microprocessor to accommodate application changes even after a product is in the field. But, the traditional system designer was experienced in logic design and hardware development and generally knew little about programming and software. This forced users to learn a new discipline to take advantage of the LSI devices. As microprocessors advance to VLSI and higher integration levels, the programming aspect is mitigated. Higher level languages make it easier for relatively inexperienced software people to take full advantage of the microprocessors. At the LSI level, Metal Oxide Semiconductor (MOS) circuits were introduced. Generally, TTL interface levels are supported to simplify usage.

The first VLSI circuits were large dynamic RAM chips. Other early VLSI circuits are floating point arithmetic circuits and signal processing circuits. These VLSI chips provide the functional equivalent of several boards of LSI and MSI logic. An important VLSI issue is the desire to use new power supply and logic signal levels. So far, users have been willing to make speed and power sacrifices so that their VLSI chips will be compatible with previous technologies. In the future, new power supply and interface standards will necessitate redevelopment of support functions that are currently taken for granted.

The VLSI Context

The design source for integrated circuits is the next important issue. As shown in Table 1-3 with SSI and MSI circuits, the basic design source was generally the modules for mainframe computers. There were extensions to take advantage of the high density of MSI which were departures from previous computer modules. The largest issue, as mentioned earlier, was whether the resulting devices would be sufficiently universal for use in a variety of different applications. Certainly, if a semiconductor company developed a set of logic modules or a family of SSI and MSI chips based on the computer modules of a particular

Integration	Design Source	Issues
SSI	Computer Modules	Universality
MSI		
LSI	Breakthrough	Testability
VLSI	Cellular Logic	

Table 1-3. Design Evolution

computer company; they knew the chips would have a good market with that computer company. The question was whether those chips would be usable in other systems or whether they would apply to only one computer line. Although many different logic families were developed in the 1960s, only a few remain viable today. The survivors are those that offered good performance and were reasonably universal. This universality was recognized and a number of semiconductor companies "second sourced" the chips of the families and worked with the original developer to create additional functions that were interoperable with the original chips.

Within LSI, many designs were extensions of MSI functions, but the design and concept of the microprocessor was a breakthrough. It was totally different from previous generations of circuits. For the first time performance was sacrificed to provide a high degree of programmability. The new emphasis was to strive for maximum flexibility by sacrificing speed and power to achieve flexibility.

VLSI began a trend back to cellular logic. This idea enjoyed some prominence in the 1950s in Automata theory for defining building blocks for Turing machines, universal computers, and self reproducing modules. Concurrently, there was a push to examine the use of arithmetic processes with a small library of primitive cells. Developing two or three cells that would snap together to build a larger function (in the same sense that pieces of a jigsaw puzzle snap together to create a picture) was envisioned. There was a very substantial sacrifice in speed and complexity for the cellular logic designs relative to custom, highly-optimized designs. The one advantage was that it was only necessary to develop a very small number of cell types. In present VLSI circuit design, minimizing the design effort by constructing a chip using replicated simple cells helps to circumvent the high design cost.

There is concern about testability at the LSI and VLSI levels of integration. VLSI circuits are often so complex that testability must be considered in the initial design phase. Even if a chip is testable, it may require impractically long test sequences. Such chips are hard for users to test and may complicate the diagnosis of problems within a system; for example, when a system fails, which chip or other part of the system has failed? The 1980s has witnessed a renewed interest in work on testability initiated in the 1950s. In the early days, the concern was with testability of large digital computers. Now single chips are as complex as early computers so that many of the same concepts apply.

1.2 The VLSI Design Process

Figure 1-2 shows how the system designer interacts with the VLSI design process. The designer develops preliminary chip specifications. These are analyzed to determine the implementation requirements; often in the form of algorithms and preliminary chip architectures. Where possible, the designer develops candidate architectures of cellular logic building blocks which are analyzed to determine the approximate complexity. An initial complexity assessment based on the complexity (i.e., gate count and memory requirements) and the required speed is used to develop an initial estimate of the proposed chip's complexity. Based on the complexity estimate, the chip feasibility is evaluated. If the estimated complexity is too low, the system application is reconsidered to determine if more of the system could be placed on the chip. If the estimated complexity is too high, it is necessary to determine if there is a lower complexity approach to implement the chip; perhaps sacrificing some performance, accuracy, etc. This process of estimating the complexity, evaluating the feasibility, and then modifying the specifications is repeated until a chip is defined that meets the complexity guidelines for the available technology. A more refined complexity estimate is made at that point using the specific characteristics of the candidate technologies. For example, in a Complimentary MOS (CMOS) implementation, it might be possible to exploit the availability of bidirectional data transfer structures to reduce the chip complexity. When technology selection is complete, the classical chip development and fabrication process begins.

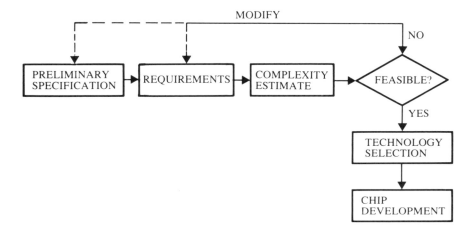

Figure 1-2. VLSI Design Process

Developing a preliminary specification and modifying it based on the available technology limits to produce a final specification are the first steps in the complete VLSI development process shown in Figure 1-3. Major steps in the development cycle include functional and gate level logic design, test design, circuit (i.e., silicon) design, layout, mask making, wafer fabrication, and final test. This book focuses on the functional, logic, and test design steps. The other development stages are effectively described in [1-3].

The benefits of VLSI are well known. Using VLSI, it becomes possible to implement a given system more efficiently. Fewer chips are used, thus the system is more reliable, lower in cost (assuming a reasonable production run), lower in power, smaller, and easier to support in the field. Alternatively, VLSI technology can be used to increase system performance while staying at a fixed complexity (i.e., cost, power, size, weight, etc.). In this scenario, VLSI implemen-

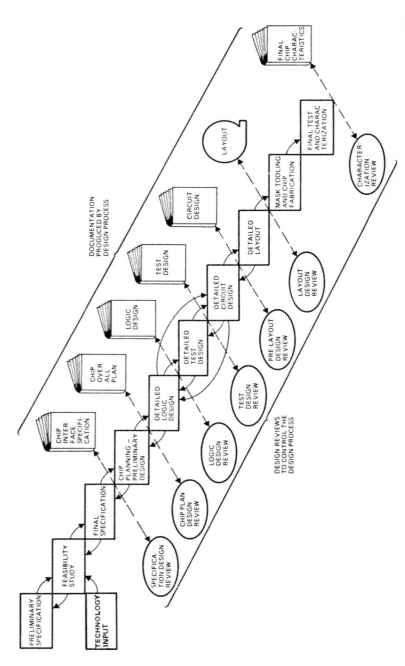

Figure 1-3. VLSI Development Process

The VLSI Context 9

tation provides better performance than conventional implementation. The performance improvements may be in the form of higher precision arithmetic, multiple elements operating in parallel to provide fault tolerance, a higher level of operating margin, etc.

1.3 Very High Speed Integrated Circuits

In 1980 the U.S. Department of Defense (DoD) initiated a major activity to develop VHSIC. The program goals are summarized in [1-4] with further details from varying perspectives in [1-5], [1-6], and [1-7]. The first program director, Larry Sumney, explains the initial accomplishments and hypotheses concerning the future of the VHSIC program and the VLSI industry in [1-8].

1.4 References

The book edited by Sze [1-3] provides an introduction to VLSI processing (i.e., crystal growth, epitaxy, photoresist, ion implementation, etc). The subscription series of [1-9] is an ongoing series that provides detailed information on various aspects ranging from VLSI processing to design and applications. The book edited by Dave Barbe [1-10] provides an overview of VLSI fundamentals. A good overview of bipolar and MOS circuit forms is included in [1-11]. Several conference/course proceedings have been published that identify current activities in VLSI design [1-12] — [1-16].

1-1 Robert N. Noyce, "Microelectronics," **Scientific American**, September 1977, pp. 62-69.

1-2 Gordon E. Moore, "VLSI: Some Fundamental Challenges," **IEEE Spectrum**, April 1979, pp. 30-37.

1-3 S.M. Sze, Ed., **VLSI Technology**, New York: McGraw-Hill Book Co., 1983.

1-4 Larry W. Sumney, "VLSI With a Vengence," **IEEE Spectrum**, April 1980, pp. 24-27.

1-5 D.F. Barbe, "VHSIC Systems and Technology," **Computer**, February 1981, pp. 13-22.

1-6   Philip J. Klass and Benjamin M. Elson, "Technical Survey: Very High Speed Integrated Circuits," **Aviation Week & Space Technology**, February 16, 1981, pp. 48-85.

1-7   Jerry Werner, "VHSIC − The Focus Shifts from Microns to Systems," **VLSI Design**, November/December 1982, pp. 22-26.

1-8   Larry W. Sumney, "VHSIC: A Status Report," **IEEE Spectrum**, December 1982, pp. 34-39.

1-9   Norman G. Einspruch, **VLSI Electronics Microstructure Science**, New York: Academic Press.

1-10 D.F. Barbe, Ed., **Very Large Scale Integration (VLSI) Fundamentals and Applications**, New York: Springer-Verlag, 1980.

1-11 M.J. Howes and D.V. Morgan, eds., **Large Scale Integration Devices, Circuits, and Systems**, New York: John Wiley & Sons, 1981.

1-12 John P. Gray, ed., **VLSI 81** (Proceedings of the First International Conference on Very Large Scale Integration, Edinburgh, August 18-21, 1981), New York: Academic Press, 1981.

1-13 H.T. Kung, Bob Sproull, and Guy Steele, eds., **VLSI Systems and Computations** (Proceedings of the Conference on VLSI Systems and Computations, Pittsburg, October 19-21, 1981), Rockville, MD: Computer Science Press, 1981.

1-14 B. Randell and P.C. Treleaven, eds., **VLSI Architecture** (Lecture Notes of the Advanced Course on VLSI Architecture, University of Bristol, July 19-30, 1982), Englewood Cliffs: Prentice/Hall International, 1983.

1-15 Randal Bryant, ed., **Third Caltech Conference on Very Large Scale Integration** (Proceedings of the Third Caltech Conference on Very Large Scale Integration, Pasadena, March 21-23, 1983), Rockville, MD: Computer Science Press, 1983.

1-16 F. Anccau and E.J. Aas, eds., **VLSI 83 VLSI Design of Digital Systems** (Proceedings of the International Conference on Very Large Scale Integration, Trondheim, Norway, August 16-19, 1983), New York: North-Holland, 1983.

CHAPTER 2.
VLSI DESIGN

This chapter addresses three important topics. The first is the use of performance metrics to compare the performance of different technologies and to estimate the complexity of specific systems. These metrics are useful in assessing the feasibility of potential VLSI circuits, since the complexity required to implement candidate circuits is easily compared with the capability of the technology.

The second topic examined in this chapter is the use of semi-custom design. This can be a very efficient way to develop (i.e., design and fabricate) small to moderate production runs of specialized circuits. Semi-custom logic includes gate arrays and standard cell logic both of which use predesigned cells from a computer design library. By use of such cells, VLSI circuits are produced faster, for less cost, and with reduced risk relative to custom design.

VLSI constraints are identified in the last two sections of this chapter. The areas of packaging and testing currently represent critical limitations on the design and use of VLSI. These areas are examined in Sections 2.3 and 2.4, respectively.

2.1 Chip Performance Metrics

In the area of the chip performance metrics three approaches will be examined. The gate-rate metric is a measure of the amount of computation that is performed on a given chip. Functional Throughput Rate (FTR) is an extension of the gate-rate metric designed to measure the amount of computation per unit area. Finally, functional performance measures computation capability in the context of specific applications. This section describes the three metrics and examines their use for technology selection.

Gate-Rate

The gate-rate measure was developed in 1978 [2-1], [2-2] as an indication of the capability of various circuit technologies, and as a measure of the amount of computation needed to implement logic functions. The gate-rate is defined as the number of two input gates multiplied by their clock rate.

For consistency in comparing different technologies, the number of equivalent two input logic gates is used as an indicator of the logic complexity. Technologies that use simple (and small) structures to perform complex logic functions achieve better performance than technologies that are restricted to gating structures with lower levels of functionality. This measure of complexity is easily evaluated for most logic functions since an N input gate is realized with N-1 two input gates. This is shown by induction: A two input function is realized with one gate. Additional gates add two inputs and one output (which uses one input of a smaller structure) for a net gain of one input. The gates can be AND, OR, NAND, or NOR gates in an arbitrary mixture.

Gate speed is a difficult parameter to assess since it is highly dependent upon loading of the outputs. It is especially difficult to get realistic measurements of the practical speed of new and emerging technologies where speeds are reported for gates operating in ring oscillator configurations with minimal loading. Such measurements do not reflect the speeds achieved when driving high fanout loads. Since the goal of the gate-rate metric is to estimate circuit performance in real systems, the maximum clock rate of circuits is used as the gate speed. This favorably weighs the technologies that are capable of operating at the highest speeds.

Gate-rate measures the amount of logical computation that is performed per unit time. The units are the number of logic transitions per second. It is most useful for computationally intensive problems such as signal processing, and special purpose systems that require much arithmetic. It can be used effectively to select between various system development options, such as: 1) to develop a very high speed single channel implementation or 2) to partition the system into multiple parallel channels operating at a proportionately lower speed.

An example of the gate rate for three different technologies is shown on Figure 2-1. The fastest technology is Gallium Arsenide. It achieves clock rates of up to about 1 GHz with up to about a thousand gates on a chip. It is generally possible to create single chip GaAs implementations of functions with requirements for up to a thousand gates and clock rates up to 1 GHz. The second technology is an advanced bipolar technology which achieves clock rates of up to about 100 megahertz. The density is limited to about 10,000 gates on a chip. Finally NMOS is shown which is capable of clock speeds of up to about 10 megahertz and densities up to about 100,000 gates on a chip.

Figure 2-1 also shows a line that corresponds to a gate-rate of 10^{12} transitions per second. Note that each of these three technologies is capable of gate-rates of 10^{12} transitions/sec if operated at the optimum clock rate. A system can operate at 10 megahertz level with NMOS, at 100 megahertz using the high density bipolar technology, or all the way up to 1 GHz with Gallium Arsenide. Many technologies (e.g., CMOS, TTL, I^2L, ECL, CMOS/SOS, etc.) were omitted from Figure 2-1 to prevent confusion. Data on the maximum gate count and the maximum clock rates are generally available from technology suppliers.

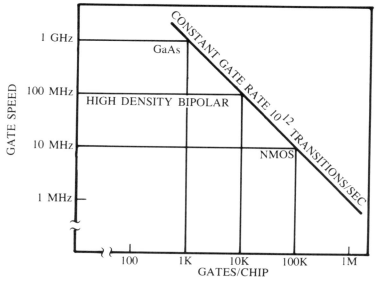

Figure 2-1. Example Gate Rates

Functional Throughput Rate (FTR)

The gate-rate is a useful metric for evaluating the computational capability of technologies but it is not an optimum measure of computational density. It is desirable to remove chip size limitations which may vary from one vendor to another. Although initially defined to be the product of the number of gates times their clock rate [2-3] (i.e., the gate-rate) for this latter application, the Functional Throughput Rate (FTR) was developed. The FTR is defined as the gate-rate achieved by a circuit divided by the chip area [2-3]. It uses the number of equivalent two input logic gates per chips the maximum clock rate of complex chips, and the area in square centimeters. FTR is expressed in units of gate-Hz/cm^2.

Functional Performance

The functional performance metric involves a measurement of the utility of a circuit at an application level. In many situations the real interest is to estimate the rate at which computation is performed. The chip count required to implement a processor divided into the processor throughput gives a measure of the throughput per chip (often expressed as Millions of Instructions Per Second, per chip).

In data processing, the processor throughput is often greatly dependent upon the mix of instructions. Performance data in the context of a mix appropriate for one application may not accurately reflect the performance for another application. It is often necessary to create an instruction mix for the specific application and estimate the throughput for this mix. If many of the operations are context dependent (where the speed of the operation varies depending on previous instructions that have been executed), it may be necessary to run benchmark programs on an Instruction Set Architecture Simulator to estimate the throughput with reasonable accuracy.

The problem is usually somewhat simpler in signal processing applications. Here processor performance is easily characterized in terms of the arithmetic operations per unit time. Similarly, VLSI circuits for signal processing can be evaluated on the basis of the rate at which they perform the appropriate arithmetic operations. For example, FIR filter implementations can be evaluated on the basis of the number of filter terms computed per unit time where evaluating each filter term requires a multiplication and an addition. Such an evaluation is used in Chapter 5 to compare four FIR filter implementations.

Similarly, FFT processor performance can be measured by the rate at which radix 2 butterflys (a standard operation consisting of four multiplications and six additions) are computed. The radix 2 butterfly computation rate is evaluated for the FFT processor case study in Chapter 6.

Use of Metrics

Metrics are really only starting points in evaluating processor architectures in the context of rapidly evolving technology. They are useful in estimating the feasibility of implementing candidate functions with specific technologies, but should not be expected to provide exact results. For example, assume a candidate function is being considered for implemention with a technology. The amount of computation required to perform the candidate function are expressed as P in the units of the metric. The best estimate of the amount of computation that can be performed by a single chip (the technology capability) is expressed as T in the units of the metric. Then a first order estimate of the number of chips, C, necessary to implement the function is given by:

$$C = P/T .$$

For example consider a sonar processor. The processor requires a gate rate (P) of 10^{14}. From Figure 2-1 a technology capability (T) gate rate of 10^{12} is assumed. Thus a first order estimate of the processor complexity is 100 chips. There is considerable uncertainty in the estimates, both in the gate-rate required to implement a function and in the attainable performance per chip. Different implementation strategies can easily change the complexity estimates by an order of magnitude, but this estimate indicates that the sonar processor complexity should fall in the range of 30 to 300 chips. The complexity estimate facilitates estimating the size, weight, power, reliability, etc. since these parameters depend on the number of chips, to a first order.

Metrics provide a useful first order estimate of the requirements, but must be used with caution. For example if a requirement exists for a gate rate of 10^{13} and available technology provides 10^{12}, as many as 100 circuits may be required. The estimates are probably only accurate to within an order of magnitude, and ignore many parameters that may be crucial for specific applications, e.g., radiation hardness, reliability, fault tolerance, or built in test. None of these issues is considered in developing the metrics but each may drastically change the capability of the technology. For example, memory capability changes markedly when the operating temperature range or radiation environment is changed. This is because more charge must be stored for stability in more demanding environments. If the chips are required to operate over wider temperature ranges the technology capability estimates must be modified accordingly. Another important issue that has not been addressed is the number of distinct chip types that need to be developed. In many signal processing applications it is possible to use a large number of identical circuits in pipeline or parallel architectures.

A final aspect of the use of metrics is the need in some situations to extend them to provide a more complete characterization of complex systems. One approach is to take either the gate rate or the functional performance metric which are oriented towards measuring computational performance and develop a similar metric for memory. Then a system is characterized by the sum of the requirements for computation and memory. For the memory characterization, it may be appropriate to simply measure the total capacity. Alternatively it may be necessary to develop a metric that gives a better picture of memory speed. A candidate metric is the bit-rate. It is the product of the number of bits stored in the memory times the access rate. Like the gate-rate, it emphasizes the external performance of the memory. Such a metric is applied to a general purpose processor by realizing that the Central Processing Unit (CPU) will have a gate-rate requirement. The memory section will have a bit-rate requirement and the control unit has requirements for both logic (gate-rate) and memory (bit-rate). The sum of the requirements for the sub components defines the total system requirements. This combined computation and memory characterization helps clarify the traditional tradeoff between logic and memory. This tradeoff applies even to signal processing. For example, in spectrum analysis via the Fourier transform the traditional Discrete Fourier Transform (DFT) algorithm requires little or no memory, but requires much computation. The Fast Fourier Transform (FFT) algorithm requires memory and a more complex control unit, but reduces the computation significantly so that if memory is relatively inexpensive, the FFT is often more efficient than the DFT.

2.2 Semi-Custom Design

Techniques that use prefabricated or predesigned logic elements to implement custom logic are referred to as semi-custom logic. Computer aids are used to connect the predeveloped components thereby producing complete circuits. There are two generic forms of semi-custom logic. The first is the gate array where an array of gates are prefabricated without interconnection between the gates. The custom interconnection of the gates is performed as a final step to implement the desired logic. Only a few custom masks are required for gate arrays since only the last few mask layers are customized for each chip design.

VLSI Design

The second approach, standard cells, is "more custom" than the gate array in that it uses a library of pipelined cell designs that are placed and connected via automated layout and routing programs. Those cells are predesigned and precharacterized logical entities (similar to MSI circuits). The chip design is developed by using the cell library to create a custom mask set. The term "standard cell" is derived from standard height cells as the height of the cells used in most standard cell design systems is constant to simplify layout. Computer aids are used to place the cells and route interconnections between the cells to implement the custom logic and to estimate the speed and power requirements. Standard cell circuits require development of a complete set of masks.

Full custom VLSI is the sizing and placement of transistors and their interconnection to achieve maximal circuit density and speed with minimum power. Full custom design require the development of a complex set of masks. A rough comparison of the cost and performance ranges of gate array, standard cell, and custom is shown in Figure 2-2. Gate array circuits provide the lowest design cost but attain lower levels of performance than circuits developed with the other technologies [2-4], [2-5]. Generally, custom chips provide the highest performance, but only after an expensive design effort. Standard cell designs fall between the other two technologies in both cost and performance capabilities.

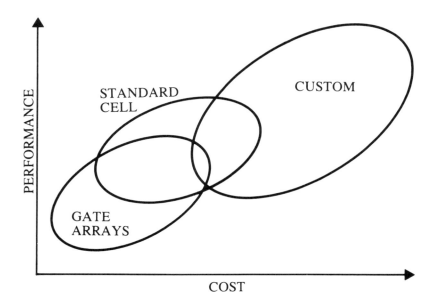

Figure 2-2. Design Technology Options

Gate Array Logic

The typical gate array development follows the process shown on Figure 2-3 (after [2-6]). Arrays consisting of circuits with standardized patterns of unconnected gates are fabricated and stockpiled as shown on the left side of the figure. An example unmetalized gate array is shown in Figure 2-4. This is a very low density (158 gate) array where the regular array of gates is more apparent than for contemporary high density arrays. The personalization (i.e., the design of the interconnection routing pattern) is designed with an automated design system. Automated routing programs provide the design for the interconnection that connects the gates selected by the automated placement program. The preprocessed wafers are customized by fabricating the interconnection pattern and the resulting circuits are packaged and tested. This approach has the advantage that only a few custom masks are required for the interconnection design instead of the typical ten to twelve masks for a custom chip and can shorten the fabrication cycle to a few weeks. It has the disadvantage that the complexity is strictly limited and physical constraints may preclude interconnection for some highly "random" logic networks.

An important issue in gate array design is the amount of area reserved for gate interconnection. If too little area is provided it may be impossible to automatically route the circuit to create the interconnection pattern. There is a tradeoff in that manual routing is more effective than automatic routing. It may be possible to use a gate array that has too little area for automated interconnection routing and still get usable results by resorting to manual routing, but this defeats much of the purpose of the gate array which is to simplify the design process. At the other extreme, if the interconnection area is too large then the resulting dice are larger than necessary, which may impact the cost. Such oversized circuits won't necessarily exhibit lower yield since flaws in the unused routing area do not cause failures, but the larger die means fewer circuits per wafer.

It is possible to estimate the routability of gate arrays. The current results are largely empirical but provide useful design guidance. The routing constant, K, is defined as the number of horizontal routing tracks, H, times the number of vertical routing tracks, V, divided by the number of equivalent two input gates on the array, G. That is:

$$K = HV/G$$

If $K < 5$, experience indicates that it will be very difficult to use an automated router for the gate array. If $K > 10$ it is generally possible to use an automated router even with fairly high gate utilizations (i.e., 80% or more). In the intermediate range $5 < K < 10$, the automated "routability" is highly dependent

VLSI Design

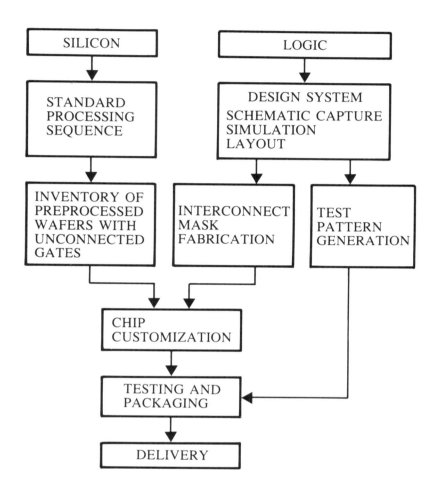

Figure 2-3. Gate Array Development (after [2-6])

Figure 2-4. Unmetalized Gate Array Chip

upon the gate utilization of each design and the optimality of the placement (i.e., the assignment of physical gates to implement the logical gates of the design). The routing constant concept applies to "average" random logic. It may be pessimistic for pipeline, or cellular structures where a regular pattern of cells can be used, or where an efficient gate placement has been achieved. Techniques for gate array routing are presented in [2-7] - [2-11].

Standard Cell Logic

Standard cell logic is an extension of the gate array concepts. Chips are built using cell designs from a library where each of the cells has been predesigned, and thoroughly characterized so that there is a detailed understanding of the behavior of that cell under different conditions. An example standard cell is shown on Figure 2-5 (after [2-12]). A typical library is similar to catalogs of SSI and MSI circuits with as many as 600 cell types. Each cell design is characterized for a variety of conditions, for example, speed vs. power for various levels of interconnection loading. The designer selects cells which are placed and routed

VLSI Design

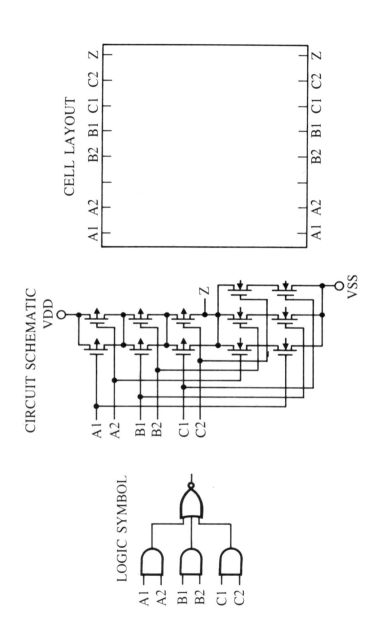

Figure 2-5. Example Standard Cell (after [2-12])

by CAD programs to implement the desired logic. In contrast to gate arrays, in standard cell designs the designer selects from a wide variety of cell types and power levels. CAD programs provide the cell placement, interconnection routing, check circuit loading, and estimate the delays. Detailed simulations verify the logical functionality and estimate the circuit delays. This normally provides a first pass to the designer who modifies the design to achieve the required speed with minimum power.

The process for the design of a typical standard circuit chip is shown on Figure 2-6. The logic design is translated into a functional design by the designer and the computer aided design system under control of the designer. Cell types and speed-power levels are selected, speeds are estimated via simulation, placement programs minimize interconnect lengths and delays along critical paths, etc. Finally, the layout system places bonding pads to create the final layout, the masks are made, the chip fabricated, and the resulting chips are tested.

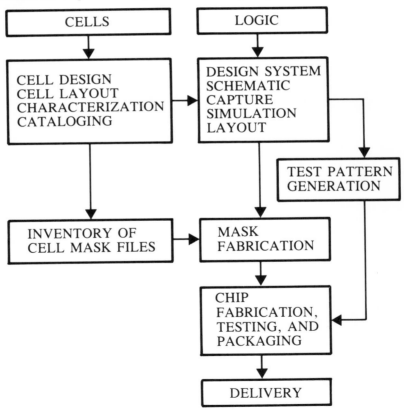

Figure 2-6. Standard Cell Development

VLSI Design   23

An example standard cell chip is shown on Figure 2-7. Eight rows of cells are placed in pairs across the chip. Since all cells are of a standard height, the cells form a uniform structure with variable spacing between rows to account for differing interconnection densities. As with the gate array of Figure 2-4, this is a very low density chip selected to clearly show the basic structure of a standard cell chip.

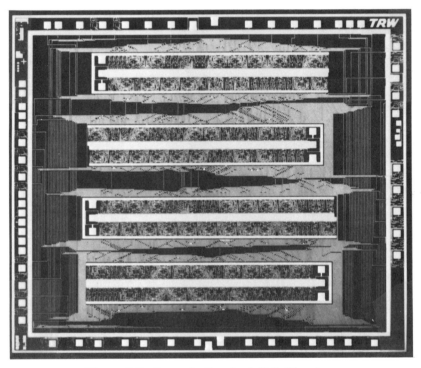

Figure 2-7. Example Standard Cell Circuit

Comparison of Approaches

The two semi-custom approaches are compared with full custom design in Table 2-1. The design time to go from a functional layout to finished chips is from one to three months for gate arrays, three to six months for standard cell chips, and a year or more for full custom chips. The design time for custom designs depends on how similar the design is to previous designs and how complex the chip is.

Technology	Design		Complexity
	Time	Cost	
Gate Array	1-3 Months	$10K to $75K	2K to 8K Gates
Standard Cell	3-6 Months	$35K to $100K	5K to 12K Gates
Full Custom	1 year	$250K	Up to 100K Gates

Table 2-1. VLSI Circuit Design Options

Typical design costs for gate arrays are in the range of $25,000 to $50,000 [2-5], [2-13]. Standard cell chips are normally about twice as expensive, whereas full custom chips generally will cost $250,000 or more. Thus gate arrays are relatively quick and inexpensive to design. Standard cell chips take twice as long to develop and are roughly twice as expensive to design and full custom take much longer to develop and circuits are much more expensive. In high quantities, the recurring cost per chip is highest for gate arrays, intermediate for standard cell chips, and lowest for custom chips.

The complexity of chips that can be developed varies from the order of 2000 gates to 8000 gates for gate arrays to roughly twice that for standard cell designs, and to as much as an order of magnitude greater for full custom designs. Currently, typical full custom chips achieve densities of 10,000 to 25,000 gates. The designer has minimal latitude in selection of gate functions and power with gate arrays, since the gates are fixed by the physical design of the array (although it is possible to program the power level of gates in some arrays). With standard cell designs, the designer must use cells that are in the library. Finally the custom designer has total freedom in cell design (which accounts for much of the cost and schedule differential between semi-custom and custom VLSI).

Figure 2-8 presents a qualitative cost/volume comparison (after [2-14]). In very small quantities the design and mask fabrication are dominant. As a result, gate arrays are cheapest, followed by standard cell circuits, and custom designs are the most expensive. The resulting dice are largest as gate arrays, intermediate as standard cell implementations, and smallest for custom designs. It is the fabrication cost that dominates for large quantities so the cost per die is lowest for custom designs, highest for gate arrays, and intermediate for standard cell circuits. Since gate array design costs are low, but recurring costs are high, the gate array is often the best choice for small quantities of chips. Standard cell designs cost more to design since all mask levels must be designed, but the resulting chip size is smaller (and the die cost is less). For moderate to large production runs, the sav-

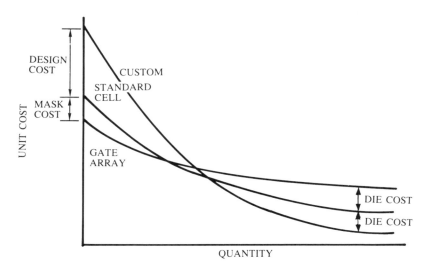

Figure 2-8. Cost/Volume Relationship for Semi Custom Technologies

ings in recurring cost more than compensates for the increased design cost. At very high quantity levels full custom designs become the most attractive from total cost viewpoint, since the recurring cost is lowest due to the small chip size.

Several caveats are important. In some situations the design is likely to be modified to accommodate changing applications. In such cases, the use of gate arrays with advanced CAD allows changes to be implemented quickly and easily (and inexpensively). Similarly, "time to market" considerations may dictate the use of semi-custom designs even though a full custom design would be more cost effective. Alternatively environmental constraints (i.e., special temperature ranges, radiation environments, etc.) or performance requirements may demand the use of standard cell [2-15] or full custom design even for relatively "simple" circuits.

2.3 Packaging Constraints

This section examines packaging, a critical constraint that interacts strongly with chip architecture. Current packaging severely limits the number of chip pinouts thereby forcing the development of architectures that require a minimum number of external connections. Packaging technology depends directly on mechanical fabrication which is evolving rather slowly. To accommodate current and planned chips, packages are needed with more pins, better cooling, and larger cavities.

Pinout Requirements

The first major analysis of the interconnection complexity of logic was performed by E.F. Rent of IBM who analyzed the external pin requirements for circuit modules used in general purpose computers in 1960. Rent discovered that the logarithm of the module pin count is roughly proportional to the number of circuit blocks that make up the module. Subsequent analysis gives the relationship $P = 4.17C^{.65}$ where P is the number of external pins for the module and C is the number of circuit blocks comprising the module. Rent's initial work was published within IBM in 1960 as reported in [2-16]. Subsequent papers have considered logic partitioning in the IBM System/360, the RCA LIMAC, and other systems.

Within the VLSI era a similar relation has been used to predict the number of external pins, P, for a VLSI circuit as a function of the number of logic gates that comprise the VLSI, C. The relation is called Rent's rule in honor of his early work in estimating pin counts:

$$P = AC^B$$

The constant A depends on the complexity of the logic gates. It is usually in the range from one-half to four or five. The exponent, B is usually in the range from 0.4 to 0.8. Figure 2-9 shows this relationship for a typical set of values (A = 1 and B = .57). As the circuit count varies between 100 and 10,000 the pin requirements range from 14 to 190.

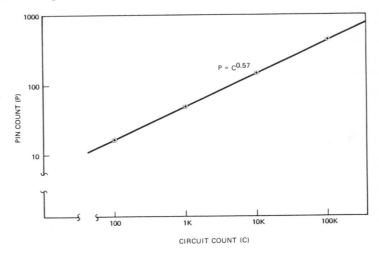

Figure 2-9. Rent's Rule Example

This model is generally accurate for circuits that are of the same basic technology over moderate complexity ranges, say an order of magnitude. It does not predict pin requirements well for new generations of technology. An obvious example of the problem is to compute the constants A and B for SSI and MSI circuits and then use the model to estimate the pin count of a large (i.e., 50,000 gate) VLSI microprocessor. The resulting estimate on the order of 500 pins fails to account for the functional synergy that occurs with large VLSI circuits. Since the VLSI circuit implements a complete function (or multiple functions) the number of pins is greatly reduced relative to lower integration level technologies where functions are implemented on several interconnected chips. As a second example, consider a hypothetical superchip technology that allows 10,000,000 high speed gates on a chip. In most cases, only very few pins would be required for the superchip because complete systems will be realized on a single superchip. As a result, only input, output, control, and diagnostic ports are needed. This is in marked contrast to the current situation where separate microprocessors, memories, controllers, I/O processors, etc. would be used to perform the same function. With the superchip the pin out problem is eliminated. This hypothetical example doesn't reduce the value of Rent's rule which has been and continues to be effective for package pin estimation for modest changes in the circuit technology.

Power Consumption

An important aspect of the interconnection problem is the high power consumption of output driver circuits. The amount of power, P, consumed by the output driver during switching is a function of the supply voltage, V_{CC}, the logic levels, V_H and V_L, the interconnect capacitance, C, and the signal slew rate, de/dt. P is given by:

$$P = \frac{V_{CC} - V_L + V_{CC} - V_H}{2} C \, de/dt$$

For example, consider a chip with TTL logic levels ($V_L = 0$, $V_H = 3$), normal printed circuit board capacitance (50 picofarads), and a switching time of 4 nsec. Since the output must change by 3 volts in 4 nsec, the slew rate is $\frac{3}{4} \times 10^9$ volts per second. The total power dissipation is about 1/7 of a watt, for single output driver. Thus a chip that has 150 pins with 100 outputs, could dissipate 13 watts for the output drivers. This is the dynamic (or switching) power, there is an additional technology dependent steady state power that has been neglected. There are several ways to mitigate this problem: reduce the speed, reduce the load capacitance, and reduce the number of outputs. If the number of outputs is reduced, it reduces the power generally and also simplifies the packaging because fewer pins are required. Reducing V_{CC} is also advantageous but results in incompatibility with existing circuits unless special power supplies (and logic level translators) are used.

Packaging Technology

The current industry standard package for commercial VLSI is the 64 pin dual-in-line package. It consists of a plastic or ceramic body with pins along the two sides. The pins are spaced on 0.1 inch centers; since there are 32 along each side the package is about 3.2 inches long. The hermetically sealed ceramic 64 pin DIP package is relatively fragile. As the package is inserted into a socket there is upward pressure on each of the pins making it very easy to break the package.

A cutaway view of an example high power dissipation 64 pin dual-in-line package is shown in Figure 2-10. The chip itself is mounted upside down within the package. This provides a short thermal path to the heat sink mounted on the top of the package. The heat sink provides about three square inches of area for heat dissipation for a 64 pin DIP.

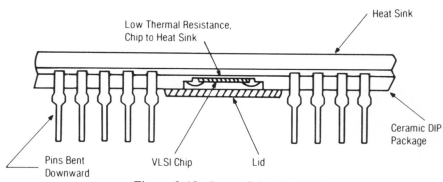

Figure 2-10. State of the Art DIP

The size of the largest available commercial dual-in-line package is shown in Figure 2-11. A linear growth model with an increase of about 4 pins/year fits very closely. Since 1976 the 64 pin package has been the largest, in the next few years there may be a package at the 84 or 100 pin level, but certainly it is not reasonable to expect dual-in-line packages with the 150-200 pin counts that would ameliorate many of the current architecture problems in the near future.

VLSI Design

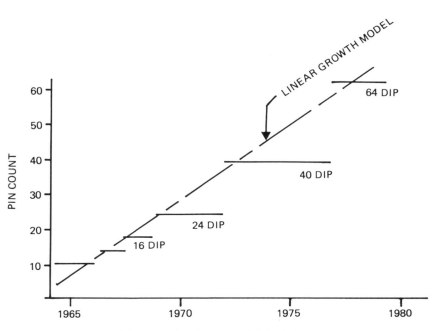

Figure 2-11. Commercial DIP Sizes

There are other packaging concepts that provide more pins [2-17] - [2-20]. Currently, there is much work in the area of leadless chip carriers (LCC). These packages are square ceramic assemblies with bonding pads around all 4 sides of the LCC. By using a pad to pad spacing of 0.05 inches and using all four sides of the carrier, the 64 pad LCC occupies only ¼ the area of a 64 pin DIP package. The pin grid array (PGA) package is a square package with a two-dimensional array of pins on a 0.1 inch 2 dimensional grid across the entire package bottom. Instead of the two rows of pins provided by the dual-in-line package, the PGA provides a matrix of pins to achieve very high packaging pin density. For 64 pins the PGA requires about the same area as the LCC, but since it uses a two-dimensional grid it becomes more efficient as the pin count grows.

2.4 Testing Constraints

Testing is another serious constraint with VLSI. It can be segmented into two major areas, design for testability [2-21], [2-22] and production circuit testing [2-23].

Design for Testability

Design for testability generally involves modifying the logic design or including additional logic to improve the controllability (i.e., the ability to establish a specific state) and the observability (i.e., the ability to deduce the current state) of the circuit. One approach for increasing the controllability and observability of logic is by the addition of intermediate storage registers within the chip. In normal operation the register behaves like a conventional latch, but upon command the contents can be accessed (i.e., loaded or unloaded) serially. This approach (sometimes called Level Sensitive Scan Design, LSSD) includes additional circuitry (sometimes called scan/set logic) that may increase circuit complexity. The complexity increase is viewed as generally an acceptable price for simplifying the testability of the chip. See for example [2-24] - [2-27] for more details.

Production Circuit Testing

The second testing constraint is production circuit testing. Here the concern is to test production circuits to verify that they operate correctly. Various testing approaches have been developed for specific classes (i.e., memory, arithmetic, control, etc.) of circuits. The remainder of this section (based on [2-28]) describes the production testing approaches used for arithmetic circuits.

There have been three "generations" of approaches to permanent fault testing: (1) non-overlapping single cell tests (2) exhaustive testing of complete circuits, and (3) exhaustive testing of overlapping subsets of the complete circuit. The advantages and disadvantages of each approach are examined in the context of parallel multiplier circuits. The goal is to verify correct functional behavior assuming that any faults are permanent (i.e., neither speed tests nor transient faults are addressed here).

Until recently most testability design and analysis of logic circuits has been based on the assumption of stuck logic faults. Such faults occur when a logic interconnection shorts to ground (stuck at 0) or to the power supply (stuck at 1). Recently however, failures involving shorts between adjacent metalization paths or between metal and underlying diffused conductors have been observed. Faults between metal and polysilicon occur as a result of breakdown of the insulating oxide. Metal to metal faults can occur as a result of flaws in the manufacturing process or as a result of metal migration. It appears likely that as the circuit density continues to increase (implying reduced interconductor separation and reduced insulator thickness) such "bridging faults" will become more prevalent.

In the discussion of testing techniques, it is convenient to use an example circuit. The circuit for an array multiplier shown in Figure 2-12 (from [2-29]) demonstrates the production testing issues. In the figure an array multiplier for forming the 14 bit product (P_{13}, P_{12}, ..., P_0) of a 6 bit number (x_5, x_4, ..., x_0) times an 8 bit number (y_7, y_6, ..., y_0) is realized with 40 full adders (shown as large circles) arranged on a regular two-dimensional grid.

In first generation testing stuck faults are assumed. For such faults, testing of complex circuits can be performed by individually testing each of the cells that comprise the circuit. Since the parallel multiplier has $n^2 - n$ cells and since each cell has three inputs (thereby requiring 8 test vectors to test), the number of test vectors is $8(n^2 - n)$. As shown on Table 2-2 for $n = 16$, 1920 test vectors are required. If this number is too high it is possible to test several cells simultaneously provided that they are far enough apart that their inputs and outputs don't interact with each other as shown on Figure 2-13.

n	Patterns	Test Time
8	448	Negligible
16	1920	Negligible
32	7936	Negligible

Table 2-2. First-Generation Test Pattern Count

This test is simple and fast, but experience shows that occasionally circuits fail for specific data patterns while passing this test. This failure of the testing scheme is apparently due to bridging faults which cause interaction between adjacent cells, which are not detected by testing each cell individually.

Second generation testing uses an exhaustive test to detect bridging faults. Production circuits are tested by comparing their outputs against known good reference circuits for all possible input patterns. This comparison test is performed at high speed with several production circuits in parallel in a simple test fixture. The test to find the known good or "gold standard" circuits can be quite slow but is done only once. For multipliers the known good circuits were found by comparing candidates against a minicomputer. For each possible data pattern the minicomputer computes the correct result and compares it with the result generated by the device under test.

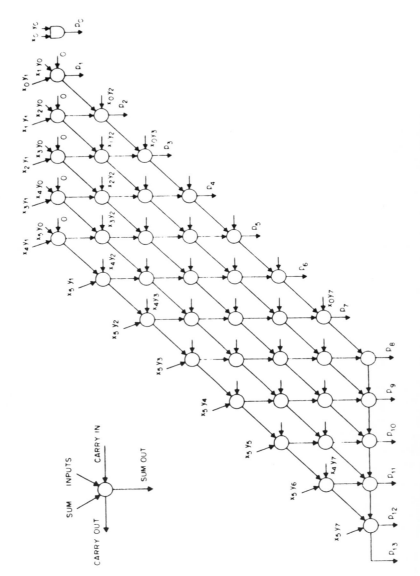

Figure 2-12. Typical Array Multiplier Block Diagram (from [2-29])

VLSI Design

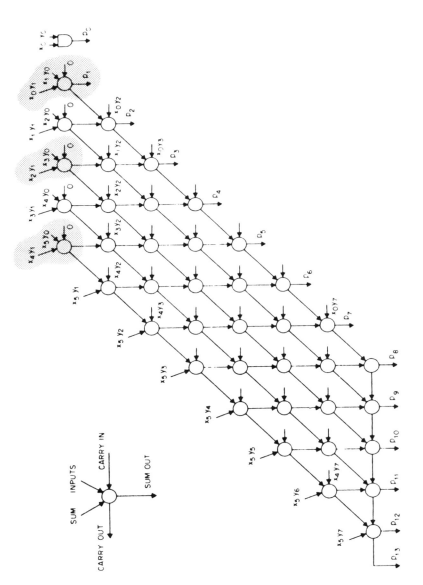

Figure 2-13. Simultaneous Testing of Independent Cells

To exhaustively test an n bit multiplier requires cycling through all possible combinations of 2n + 1 input lines (n for each of the two operands and one for the round control). The number of patterns is 2^{2n+1}, for n = 16 this is about 8.6×10^9. At a 3 micro second cycle time, the computer test to find the "gold standard" circuits takes nearly 8 hours. Even at a 100 nsec cycle time, the production test takes 14 minutes to complete as shown on Table 2-3. Fortunately such production tests can be performed on many devices in parallel. Exhaustive tests are currently used primarily with small word size arithmetic elements (i.e., 8 bits or less). Clearly as data word sizes increase the exhaustive test becomes impractical. For 32 bit arithmetic chips, the number of test patterns is in excess of 10^{19}, so that the test requires 117,000 years at a 100 nsec test cycle time!

Third generation testing involves a compromise between two unacceptable alternatives: a non-overlapping cellular test pattern that passes some defective circuits or an exhaustive test that takes too long. A potential solution is to exhaustively test sub arrays consisting of overlapping groups of cells. A two dimensional array of m^2 cells (m cells vertically by m cells horizontally) has 2m inputs so it requires 2^{2m} test patterns for exhaustive testing. Dividing the array into quarters produces four sub arrays consisting of m/2 rows of m/2 cells each. Since each sub array is exhaustively tested by 2^m test patterns, testing the total array requires 4×2^m test patterns. This array division process can be repeated until each cell is tested individually thereby producing the first generation single cell test (that failed to detect bridging faults). The minimum test which detects intercell faults is a 2 by 2 sub array. On an m × m array there are $(m-1)^2$ sub array positions for a total test complexity which grows in proportion to m^2 (in contrast to growing in proportion to 2^{2m} for the exhaustive test).

n	Patterns	Test Time*
8	1.3×10^5	13 Msec
16	8.6×10^9	14 Minutes
32	3.7×10^{19}	117,000 Years

*Test time assumes 100 nsec/test.

Table 2-3. Second-Generation Test Pattern Count

For the example parallel multiplier, a standard cellular test of $8(n^2 - n)$ patterns is used. Then the cellular test is applied to all contiguous 2 × 2 cell blocks. As shown on Figure 2-14, a 2 × 2 cell block has 9 inputs. Since an n bit multiplier will have $n^2 - 3n + 2$ contiguous 2 × 2 cell blocks, a 16 bit multiplier is tested

VLSI Design

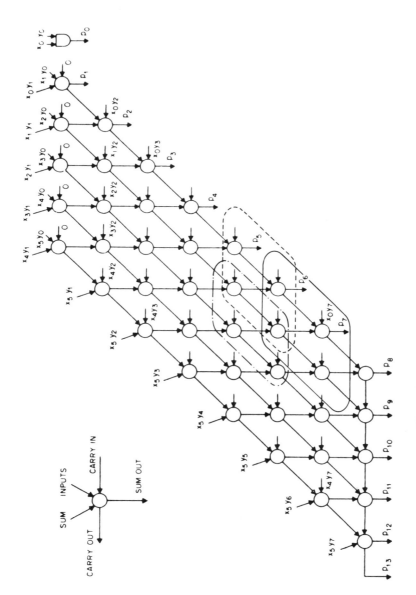

Figure 2-14. Overlapping Subset Testing

with 108,000 data vectors using the 2 × 2 cellular block test as shown on Table 2-4. As with the single cell test it is possible to test several cells simultaneously by selecting geometrically dispersed sub arrays. Although more than the 1920 patterns of a standard cellular test, this is still a quick test, which does not grown unreasonably as the word size grows. Only 476,000 patterns are required for a 32 bit multiplier. This approach detects all faults where signals in one cell are perturbing an adjacent cell. It may be extended to larger groups (i.e., a 3 by 3 block) although the number of tests increases rapidly with increasing sub array size (e.g., a 3 by 3 multiplier sub array has 17 inputs necessitating over 250,000 test patterns per block). It is expected that this approach will detect pattern sensitive circuits that pass the non-overlapping test. These "cellular" approaches where a circuit is divided into small subsections for testing appear to represent the only viable approach to testing VLSI and future larger circuits.

n	Patterns	Test Time
8	21,952	Negligible
16	109,440	Negligible
32	484,096	Negligible

Table 2-4. Third-Generation Test Pattern Count

2.5 References

2-1   Earl E. Swartzlander, Jr., "VLSI Technology for Signal Processing," **Proceedings of the Government Microcircuit Applications Conference (GOMAC),** Vol. 7, 1978, pp. 76-79.

2-2   Earl E. Swartzlander, Jr., "VLSI Architecture": in D.F. Barbe, Ed., **VLSI Fundamentals and Applications,** New York: Springer-Verlag, 1980, pp. 178-221.

2-3   Larry W. Sumney, "VHSIC: A Status Report," **IEEE Spectrum,** December 1982, pp. 34-39.

2-4   Robert L. Pritchard, "Cost and Availability of Gate Arrays and Standard Cells," **VLSI Design,** May 1984, pp. 51-58.

2-5   Roderic Beresford, "Comparing Gate Arrays and Standard-Cell ICs," **VLSI Design,** December 1983, pp. 30-36.

2-6   P.J. Hicks, "Structure of Semi-Custom Integrated Circuits," Ch. 4 in P.J. Hicks, ed., **Semi-custom IC Design and VLSI,** London: Peter Peregrinus Ltd., 1983, pp. 41-57.

2-7   Michael Burstein and Richard Pelavin, "Hierarchical Wire Routing," **IEEE Transactions on Computer-Aided Design,** Vol. CAD-2, 1983, pp. 223-234.

2-8   Benjamin S. Ting and Bou Nin Tien, "Routing Techniques for Gate Array," **IEEE Transactions on Computer-Aided Design,** Vol. CAD-2, 1983, pp. 301-312.

2-9   Shuji Tsukiyama, et al., "A New Global Router for Gate Array LSI," **IEEE Transactions on Computer-Aided Design,** Vol. CAD-2, 1983, pp. 313-321.

2-10 Jeong-Tyng Li and Malgorzata Marek-Sadowska, "Global Routing for Gate Array," **IEEE Transactions on Computer-Aided Design,** Vol. CAD-3, 1984, pp. 298-307.

2-11 Ash M. Patel, Norman L. Soong, and Robert K. Koin, "Hierarchical VLSI Routing - An Approximate Routing Procedure," **IEEE Transactions on Computer-Aided Design,** Vol. CAD-4, 1985, pp. 121-126.

2-12 A.J. Kessler and A. Ganesan, "Standard Cell VLSI Design: A Tutorial," **IEEE Circuits and Devices Magazine,** Vol. 1, January 1985, pp. 17-34.

2-13 Howard K. Dicken, "Calculating the Manufacturing Costs of Gate Arrays," **VLSI Design,** December 1983, pp. 51-55

2-14 J.R. Grierson, "Selection of Semi-Custom Technique, Supplier and Design Route," Ch. 5 in P.H. Hicks, ed., **Semi-Custom IC Design and VLSI,** London: Peter Peregrinus Ltd., 1983, pp. 58-72.

2-15 Dung Nguyen, "CMOS Standard Cell Implementation for Radiation Hardness," **VLSI Design,** December 1984, pp. 96-101.

2-16 Bernard S. Landman and Roy L. Russo, "On a Pin Versus Block Relationship for Partitions of Logic Graphs," **IEEE Transactions on Computers,** Vol. C-20, 1971, pp. 1469-1479.

2-17 Ralph Parris and John A. Nelson, "Practical Considerations in VLSI Packaging," **VLSI Design,** November/December 1982, pp. 44-49.

2-18 John W. Balde and Don Brown, "Alternatives in VLSI Packaging," **VLSI Design,** December 1983, pp. 23-29.

2-19 Willard Booth, "VLSI Era Packaging," **VLSI Design,** December 1984, pp. 22-35.

2-20 William W. Staley, "The Impact of VLSI on Electronic Packaging," **VLSI Design,** March 1985, pp. 62-63.

2-21 T.W. Williams, "Design for Testability," in P. Antognetti, D.O. Pederson, and H. de Man, eds., **Computer Design Aids for VLSI Circuits,** Boston: Martinus Nijhoff, 1984, pp. 359-416.

2-22 Thomas W. Williams and Kenneth P. Parker, "Design for Testability — A Survey," **Proceedings of the IEEE,** Vol. 71, 1983, pp. 98-112.

2-23 Eugene R. Hnatek and Beau R. Wilson, "Practical Consideration in Testing Semicustom and Custom ICs," **VLSI Design,** March 1985, pp. 20-42.

2-24 Frederick P. Beucler and Michael J. Manner, "HILDO: The Highly Integrated Logic Device Observer," **VLSI Design,** June 1984, pp. 88-96.

2-25 E.J. McCluskey, "A Survey of Design for Testability Scan Techniques," **VLSI Design,** December 1984, pp. 38-61.

2-26 Shared C. Seth and Vishwani D. Agrawal, "Cutting Chip-Testing Costs," **IEEE Spectrum,** April 1985, pp. 38-45.

2-27 G. Grassl, "Design for Testability," in Paul G. Jespers, Carlo H. Sequin, and Fernand van de Wiele, eds., **Design Methodologies for VLSI Circuits,** Rockville, MD: Sijthoff and Noordhoff, 1982, pp. 249-284.

2-28 Earl E. Swartzlander, Jr., John A. Eldon, and De D. Hsu, "VLSI Testing: A Decade of Experience" **COMPCON Proceedings,** Spring 1985, pp. 392-395.

2-29 Lawrence R. Rabiner and Bernard Gold, **Theory and Application of Digital Signal Processing,** Englewood Cliffs: Prentice-Hall, Inc., 1975, p. 519.

CHAPTER 3.
VLSI ARCHITECTURE

This chapter examines circuit architectures that are appropriate for VLSI. The VLSI design effort is a major problem which can be at last partially mitigated by applying cellular logic design concepts. The resulting chips, while perhaps not the fastest or simplest, are extremely efficient in terms of achieving high performance with reasonable design effort. This chapter also addresses interface concepts, since clever interfaces can reduce the number of package pins without sacrificing performance or flexibility. The architecture development of a parallel multiplier is examined to demonstrate the application of these points. Finally, other functions are examined to indicate the circuit architecture breadth.

3.1 Chip Design

In the chip design and layout process the basic problem is complexity. Current chips frequently have 10,000 gates, some chips are approaching 100,000 gates and the trend suggests 1,000,000 or more gates within the next decade. The design effort grows at least as fast as the gate count. Although much progress has been made in Computer Aided Design (CAD), most systems require much human intervention to produce good chip designs. It is currently estimated that the design cost for VLSI chips is approximately $100 a gate [3-1] [3-2]. This suggests that to design a million gate chip will be very expensive unless cost reducing approaches are developed. [3-3].

One approach to reducing the design cost is cellular logic. The idea is to design a few cell types and replicate them in one, two, or more dimensions to realize the desired function. This approach is widely used in the memory field where large memory chips are developed by designing a memory cell which is then replicated in two dimensions to produce the memory array. Cellular logic can be applied at the processor architecture level [3-4]. It applies in part to arithmetic because most arithmetic algorithms are based on approaches developed for human use. People generally use simple algorithms which are easily implemented with celluar logic. Consider, for example, multiplication. The basic multiplication approach used is to form a matrix of partial products and then sum the elements in each column. In VLSI, multiplication is implemented by developing a cell that forms a bit product and sums it with other bit products in the same column to produce sum.

Figure 3-1 shows the functional configuration of a 16 × 16 bit parallel multiplier, which demonstrates the small number of cell types that are needed to create such a structure. At the center is the multiplier array which is implemented with one cell type that is replicated n by n-1 times in two dimensions. The interfaces are implemented with a register "slice" that is replicated n times to make each n bit register. This same cell is used for both the input registers and the output registers. For the output a third cell type is used to implement the output

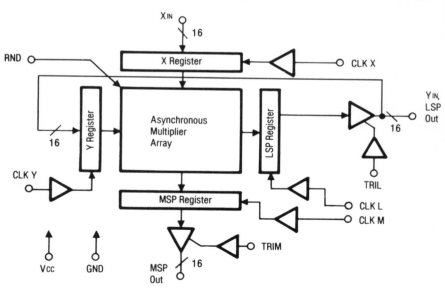

Figure 3-1. Parallel Multiplier Block Diagram

drivers. This suggests that only three cell types are required. In practice the situation is generally more complex. For example one of the operands may be multiplexed on the same pins as the least significant half of the product, so a modified input cell and a modified output driver cell will be required giving a total of five cell types. This means that the several thousand gate multiplier function can be designed by designing only five cells and their interconnection. If the average cell complexity is 20 gates, the total circuit design task is 100 gates (a savings in design effort of a couple of orders of magnitude). The design simplification is not quite as great as the gate ratio indicates since the building block cells have additional constraints to insure that all interface lines are correctly placed, but it is a significant improvement.

The multiplication array shown in Figure 3-2 for an 8×8 multiplier [3-5] indicates the approach used for most array multipliers. It is constructed as an n by n-1 array of full adder cells. Each cell has three inputs: one that is the logical AND of one bit from each operand and two from other cells. The cell forms two outputs, the sum and carry. This one-cell type is replicated 56 times to implement an 8×8 multiplier (240 times to implement a 16×16 multiplier). Again, there are some slight complications. Along the top row and at the lower right corner half adders are used. Since a half adder can be realized as a full adder with one input grounded, it is possible to implement the array by designing one basic cell which is modified slightly for positions in the array. Figure 3-3 shows the resulting 16 bit multiplier. There is a central array consisting of the two dimensional array of adder cells with register (and I/O) cells around the edges.

Beyond the advantage of reduced design complexity, this approach facilitates the development of similar functions. For example, Figure 3-4 shows a multiplier accumulator where the basic multiplier structure is augmented by adding an accumulator (i.e., an adder with a feedback loop from the output register to the input) to accumulate sums of products. This design uses the existing multiplier structure with an accumulator cell inserted between the multiplication array and the output register. Here a design for a several thousand gate chip consisted of designing the accumulator cell and inserting it into the chip layout. The multiplier accumulator chip is shown in Figure 3-5. The structure looks very similar to the multiplier of Figure 3-3, and, except for the accumulator, the same cells are used. Thus the difficult problem of integrated circuit design can be mitigated through application of cellular design techniques. Hierarchical cellular design techniques with appropriate CAD support will facilitate the design of wafer scale circuits with complexities exceeding 10 million gates per wafer in the next decade.

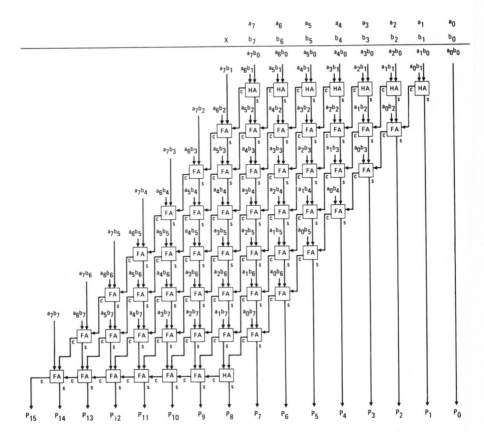

Figure 3-2. Logic Diagram of 8 x 8 Bit Parallel Multiplier

VLSI Architecture

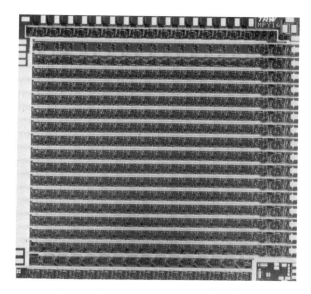

Figure 3-3. Parallel Multiplier Chip

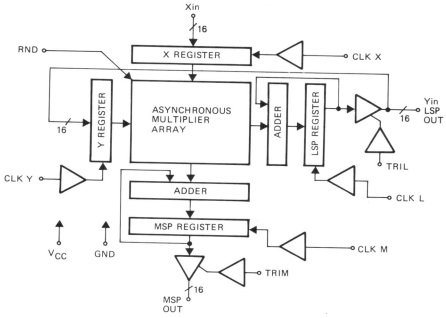

Figure 3-4. Multiplier Accumulator Block Diagram

Figure 3-5. Multiplier Accumulator Chip

3.2 Chip Interfaces

There are three approaches to interfacing (Figure 3-6). The first approach uses a fully parallel structure where all functional interfaces are brought out through package pins. The second approach uses a multiplexer where two or more of the internal ports share a single package port that switches between the internal ports as appropriate. Another approach uses serial-to-parallel converters so that a single serial line is used for each input port, with parallel-to-serial converters that drive a single output for each port. Each of these approaches is briefly examined.

Parallel Interfaces

With parallel interfaces all data connections to the circuit are connected to separate package pins. Parallel interfaces are the easiest to use since all data are readily available. This approach is very flexible as users may access all of the data or any subset by connecting the pins as appropriate. It provides the highest data rates, but uses more output drivers, a larger chip size, and more power than the other approaches.

VLSI Architecture

PARALLEL: ALL DATA EXTERNALIZED

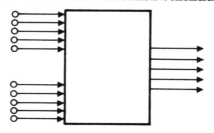

MULTIPLEXED: TWO OR MORE DATA SHARE A SINGLE PORT

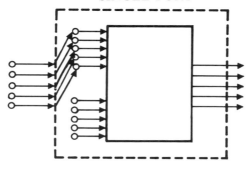

SERIAL PROTOCOLS: WORD, BYTE, OR BIT WIDE PORTS

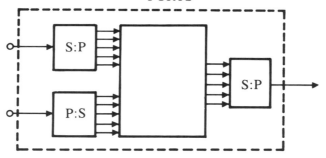

Figure 3-6. Interface Concepts

Multiplexing

The second approach uses multiplexers so that a single package pin is connected to multiple data ports. Figure 3-6 shows an input multiplexer connecting a single set of package pins to two input data ports. Section 3.3 discusses how this approach can be transparent to the user. There are disadvantages in that multiplexing may constrain the application flexibility. It reduces the maximum data rates because certain internal ports time-share a single set of external pins. In some cases, adding multiplexing means that lines run across the chip creating a more complex chip layout.

Serial Protocols

Serial protocols may be viewed as an extreme form of multiplexing where a single package pin is time-shared to carry all of the bits of each data word. Internally to the chip, serial-to-parallel converters generate parallel data for the circuit and parallel-to-serial converters generate serial outputs. Serial interfaces are attractive because they minimize package pins. As fiber optic interfaces become practical for interconnecting chips, serial interfaces will be necessary since fiber optic transducers (i.e., sensors and laser diodes) may be so large and power consumptive that using a minimal number of package "lines" will be necessary. Although fiber optics may force serious consideration of serial interfaces, currently there is no accepted standard protocol. As a result, to implement a system where serial interfaces are used in part of the system, development of either a programmable translator or a complete chip family is necessary. Serial interfaces may severely constrain the I/O data rates and thereby limit chip performance. Serial interfaces are vulnerable to a single driver failure since if the output driver fails all data is lost. In contrast, for a parallel structure, if one of the output drivers fails, it may be possible to operate in a degraded mode. Though this may be a disadvantage, it is not entirely negative. There are instances when it is useful to recognize immediately that the chip has failed. If the single serial driver fails the failure is obvious: nothing comes out of the chip. In the parallel interface, the system may continue operating for a long time before failure detection. It is relatively easy to add redundancy to serial systems since duplication of the interfaces requires adding only a few lines.

3.3 Parallel Multiplier Design Example

This section examines the design of a parallel multiplier and shows the application of the ideas presented in Sections 3.1 and 3.2. The two primary issues are the basic architecture and the circumvention of the package pin limitations.

VLSI Architecture

Basic Architecture

Figure 3-7 shows the fundamental structure for arithmetic devices. Incoming data goes into data registers where it drives directly to the arithmetic function. Either combinatorial or sequential algorithms may be used. The result is loaded into a register coupled directly through a three state driver to the package pins. A 16-bit multiplier consists of approximately 5000 two-input gates. Since current commercial multipliers operate at a 10 MHz clock rate, they achieve a gate-rate of 5×10^{10}. For the functional performance metric it computes 10^7 multiplications per second. This structure is used efficiently for arithmetic functions.

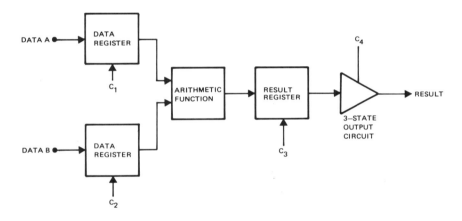

Figure 3-7. Arithmetic Function Architecture

Package Pin Limitations

When the first single chip 16×16 parallel multipliers were developed, the largest available packages had 64 pins. A 16 bit multiplier has two 16 bit input ports and a double precision output, requiring 32 pins, as well as about 10 additional pins for control, power, and ground. This totals 74 pins which exceeds the 64 pin package capability. The approach selected to solve the pin problem is to allocate separate input ports to the operands and an output port for the most significant half of the product. The least significant half of the product is multiplexed onto one of the input ports as shown in Figure 3-8. This fits easily within the 64 pin package constraint. Whether this approach is sufficiently transparent to satisfy a wide variety of applications is an important question.

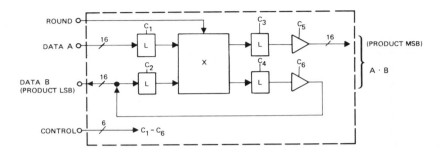

Figure 3-8. 16 Bit Multiplier Pin Constraints

This architecture has been demonstrated to be useful for many applications in both signal processing and data processing. General purpose data processing involves single port operation shown in Figure 3-9. The integrated circuit is connected to a bus that is part of a general purpose computer. The function is performed by putting the first operand onto the bus and then commanding the circuit to latch that first operand into an input register. The process is repeated for a second operand. Time is allowed for the function evaluation and the results are latched into the output registers. For the multiplier, the two halves of the double precision product are read out by putting them onto the bus on successive bus times and taking them off; either putting them back into memory or sending them to some other part of the system. For this application, the least significant product half is coupled onto the data bus so the multiplexer structure imposes no penalty. In fact, this approach could be extended to the development of single data port architectures, but the resulting circuits would only be useful for data processing applications.

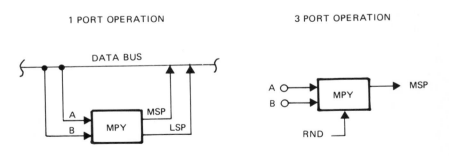

Figure 3-9. Interface Versatility

VLSI Architecture

Three port operation frequently arises in signal processing where operands arrive from two different sources. One may be an input data source, while the other is a table of kernel values for a filter, etc. The arithmetic operation is performed producing a properly rounded single precision result that can be used for subsequent processing. Since only the rounded most significant half of the product is of interest, the least significant half of the product is ignored. Thus, the multiplexer structure is truly transparent for these two application classes. In fact, a decade of commercial experience indicates that this architecture is nearly universally satisfactory for arithmetic functions.

3.4 Other Functions

Other functions that frequently arise where architectural, issues are prevalent are data conversion (i.e., analog-to-digital conversion), memory, and microprogrammed control. Each of these areas are briefly examined.

Analog-To-Digital Conversion

An especially important function in signal processing is Analog-to-Digital (A/D) conversion. Figure 3-10 shows a fully parallel A/D converter. It operates with $2^n - 1$ parallel comparators running concurrently to digitize an n bit word. The analog input voltage is applied in parallel to the positive input of each of the comparators. The reference input for the comparators is $2^n - 1$ voltage thresholds evenly spaced over the dynamic range of the A/D converter. The positive input of the comparators up to some point exceed their reference causing their output to be positive. The comparators above that point have reference inputs that exceed the analog input so they generate negative output signals. An encoder determines the "height" of the transition point and generates the corresponding digital output. This is a simple structure that lends itself very nicely to VLSI since there is only cell type. The cell includes the resistor, the comparator, and the 2^n to n encoder which is merely a ROM. To design the chip $2^n - 1$ of these cells are connected in parallel.

Figure 3-11 shows an 8 bit A/D converter. The row of 255 cells is folded several times producing a serpentine layout with a nearly square chip. From a packaging perspective, this is a great function for VLSI because it requires only a few pins: the analog input, the reference voltages, the eight digital outputs, and a few control lines.

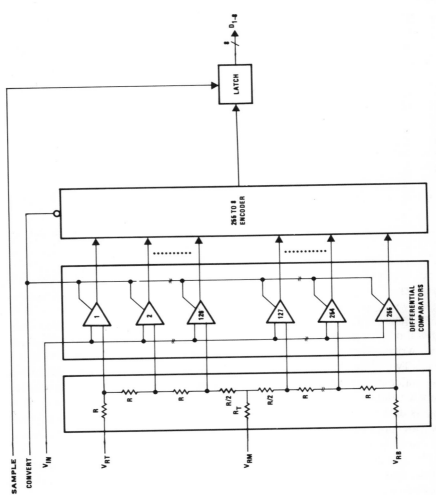

Figure 3-10. Fully Parallel (Flash) Analog-to-Digital Converter

VLSI Architecture 51

Figure 3-11. 8 Bit Analog-to-Digital Converter Chip

Memory

In signal processing, there is a strong need for specialized memory. One such special purpose memory is a multiport memory. That allows multiple data to be read/written simultaneously. The multiport memory structure is shown in Figure 3-12. Two write ports, A and B, are shown on the upper left. Each accepts a data value and an address and causes the data to be written into the corresponding location in the memory. Similarly, there are two read ports, C and D, that access data from any two memory words. There can be a problem if different data are simultaneously written into the same memory location. In most cases, the memory will be used in configurations that prevent such conflicts.

The primary application of the multipart memory is the implementation of specialized arithmetic processors. The basic approach is shown in Figure 3-13. Any of a wide variety of arithmetic functions (i.e., multipliers, microprocessors, FFT butterfly elements, etc.) are coupled to the memory so that two data are read from the memory, processed by the arithmetic function, and the result(s) written into the memory. Although three clock cycles are required to process a pair of data, the process continues with a new data set on each clock cycle.

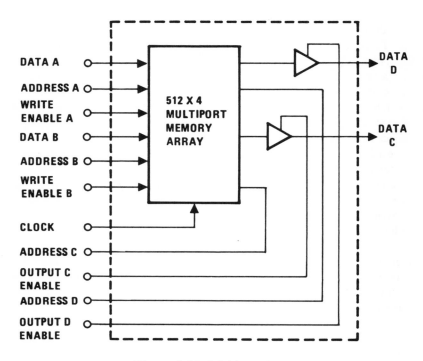

Figure 3-12. Multiport Memory

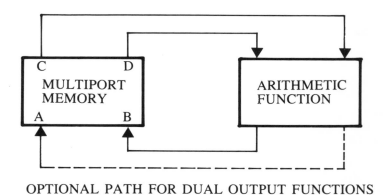

OPTIONAL PATH FOR DUAL OUTPUT FUNCTIONS

Figure 3-13 Computation Using the Multiport Memory

VLSI Architecture

Since arithmetic functions are widely available for 10MHz clock rates, performing this function with a commercial single data port memory would require a very fast memory cycle time of 20 to 30 nsec. The multiport memory implementation operates at a conservative 100 nsec cycle time.

Another multiport memory application is to serve as a buffer between processes that operate on different clocks. A radar system where data is sampled from the receiver at a very high rate (during the short period of time corresponding to the ranges of interest) and where no more data is sampled until the next range of interest is a good example. In this situation, data is acquired rapidly during short intervals, there is a dead period, more rapid sampling, another dead period, etc. The processor that operates on this data uses the entire time-period between samples to process the data. An elastic buffer is used to store incoming data and then provide the data to the processor as required over the processing time in the order it was received.

Other applications of the elastic buffer arise whenever two different clock systems are coupled. Even if the source data rate exactly matches the absorption rate, a small buffer may be used to maintain data synchronization in the presence of phase shift between the clocks.

The elastic shift register is implemented with a two-port memory, as shown in Figure 3-14. One read port and one write port are used. A counter generates the write address. Data is written in order into the memory and an independent counter generates the read address so that the data is read in sequence.

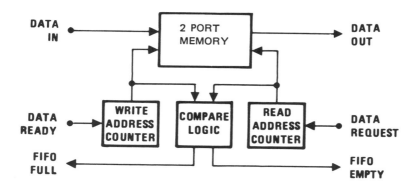

Figure 3-14. FIFO Buffer Implemented with the Multiport Memory

An example multiport memory chip is shown in Figure 3-15. The cellular structure is evident. Because this particular memory was built using a four port memory cell, the cell is relatively large and the memory capacity is only 1K bits. The VHSIC program is developing larger multiple port memories.

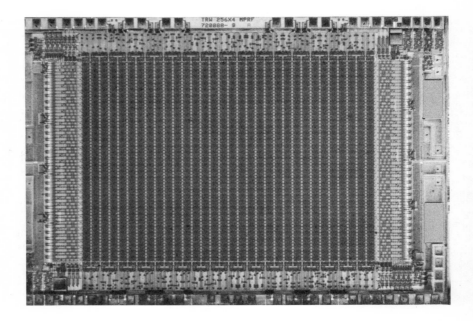

Figure 3-15. Multiport Memory Chip

*Microprogrammed Control**

Several microprogramming techniques suitable for controlling specialized processors have been developed by restricting the next state function of a classical Moore machine [3-7]. The concept is well-matched to applications which do not require complex branching logic, and is efficiently implemented with commercially available integrated circuits. The design of an example controller illustrates this concept.

---

*This section is based on [3-6].

VLSI Architecture

Traditionally, control functions have been difficult to implement effficiently using VLSI circuits. Wilkes' initial paper on microprogramming [3-8] offers an approach to alleviate this situation. The goal is to replace the "random" logic required to implement specialized control units with stored data patterns in read only memories (ROM). Programmable logic arrays have been developed and used with some success but are generally considered to be inefficient for the implementation of "ill conditioned" logic [3-9].

Attention is focused on the implementation of controllers for systems which involve the execution of well defined instruction sequences with little dependent branching, and where complex branch logic and multilevel interrupt nesting are not required. These conditions are often satisfied by digital signal processors. Signal processing frequently employs parallel processing where many identical processors operate on different data. This imposes the requirement of controller simplicity (i.e., low chip count, low power, etc.), since each processor requires a stand-alone capability for improved reliability, failure diagnosis, and fault isolation, and to provide "graceful" degradation in the event of partial failure.

A family of controller concepts is described here which, although lacking the flexibility and generality of conventional controllers, is well matched to the requirements of many special purpose processors. Specifically, these controllers provide varying degrees of data dependent next state selection (although in all cases the next state selection is quite limited by comparison to conventional controllers). These approaches may be extended to implement subroutining while retaining a low cost structure which is amenable to VLSI implementation.

In the classical Moore machine [3-7], the next state, Q_{n+1}, is a function α, of the present state, Q_n, and the present input, X_n

$$Q_{n+1} = \alpha(Q_n, X_n)$$

where Q_n is an N element binary vector and X is an M element binary vector. Note that all M input bits are considered in selecting the next state. Thus, if multiple inputs change during a state, all will impact the next state selection. For each input combination any machine state can be selected (i.e., without restricting the states to be sequential). The output, Y_n, is a K element binary vector, which is a function, β, of the present state

$$Y_n = \beta(Q_n)$$

The Moore machine is implemented as shown in Figure 3-16. On each clock pulse the present state, Q_n, is latched so that it and the next input determine the next state. Note that the next state function, α, is a 2^N by 2^M table, where each entry is an N bit number denoting the next state. The total size is $N2^N2^M$ bits, where N is the number of bits in the state and M is the number of bits in the input word. The output function, β, is a K by 2^N table.

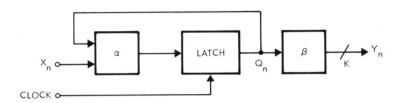

Figure 3-16. Moore Machine

Here the primary concern is to develop implementations which retain much of the flexibility of the basic Moore machine, but which use less memory. A number of methods have been used with success. These range from a simple state counter to multiplexer based systems which select between various prestored next states. Although the counter schemes are simpler to implement with commercially available circuits, the multiplexer approach is more flexible.

The simplest counter based controller is shown in Figure 3-17. The present state Q_n, is the output of a simple counter. The next state, Q_{n+1}, is either a repeat of the present state (i.e., Q_n) or the present state incremented (i.e., Q_n+1), depending on the state of **one of** the input variables (if the counter overflows $Q_n+1=0$). The input variable which controls the next state is selected by the δ field of the microprogram. If the input variable selected in state n, denoted x_δ, is true, the state counter is incremented; otherwise the state remains fixed:

$$Q_{n+1} = x'_\delta (Q_n) + x_\delta (Q_n+1)$$

If $x'_\delta = 1$ then $x'_\delta = 0$ and the next state is Q_n+1. Here the $N2^N2^M$ bit next state table of the basic Moore machine has been reduced to $\log (M) \times 2^N$ bits by restricting the next state to either the present state or the present state incremented. It is assumed throughout that commercial integrated circuit multiplexers will be used and, as a result, that M will be a power of two. In general, the field of the microprogram will be [log (M)] bits wide, where [Y] denotes the smallest integer greater than or equal to Y. The selection is also simplified since only a single input line is monitored for each state.

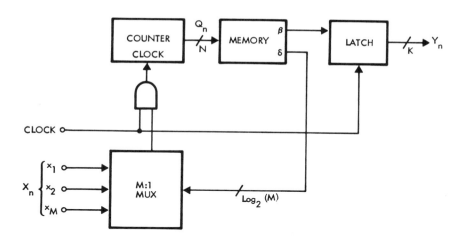

Figure 3-17. Counter Based
Microprogrammed Control Unit

The next level of controller generality is obtained with the network of Figure 3-18. Here the present state, Q_n, is the output of a presettable counter. The next state, Q_{n+1}, is either the present state incremented, i.e., Q_n+1 or an arbitrary function, γ, of the present state. This approach is sometimes called "implicit" sequencing [3-10] since the ordering of the microinstructions is implicitly the same as their execution order, except for program jumps.

$$Qn+1 = x'_\delta (Q_n + 1) + x_\delta (\gamma(Q_n))$$

The next state function requires an $(N + \log (M)) \times 2^N$ bit table, which is N bits wider than the microprogram table for the simple counter based controller, but is still much smaller than that required for the Moore machine. Presettable counters suitable for this application are commercially available in most integrated circuit families (e.g., CMOS 4029, TTL 74154, ECL 100136, etc.).

The control sequence generated by this controller is much like a low level computer program. Only a simple test and jump capability is provided (i.e., the only instruction which modifies the instruction sequence tests a single Boolean variable: if the variable is true, the program jumps to a new address, otherwise the program continues executing in sequence). To jump to either of two distinct locations, based on the state of a single input (e.g., from State 12, if x_1 is true, go to State 7; if x_1 is false, go to State 23), two instructions are required; the first is a conditional jump (to State 7) and the second is an unconditional jump (to

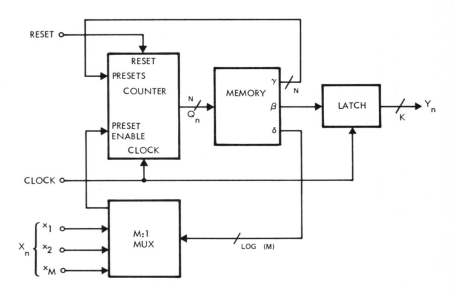

Figure 3-18. Presettable Counter
Based Microprogrammed Control Unit

State 23) which is implemented by reserving an input which is always true. Superficially, it might seem necessary to provide an input which is always false to permit sequential operation, but this can be accomplished by setting the next γ, to Q_n+1 so that if the jump is taken or not, the next state is Q_n+1.

If the counter is replaced with a multiplexer to select from two (or more) next state fields, the control unit shown in Figure 3-19 results. Here, a multiplexer selects between two next address fields, γ and ν, which are stored in the microprogram memory. As in the counter-based controllers, the δ field in the microprogram selects an input variable which determines the next state. The next stage equation for this controller is

$$Q_{n+1} = x'_\delta \, (\nu(Q_n)) + x_\delta \, (\gamma(Q_n))$$

The microprogram word for this controller requires four fields: one for input selection (δ), two next addresses (ν and γ), and the output function (β). The total size of the next state portion (i.e., δ, γ, and ν) is $(2N + \log (M)) \times 2^N$ bits. This controller allows the next state to be selected arbitrarily from anywhere within the microprogram. This generality is obtained at the cost of wider microprogram words.

VLSI Architecture

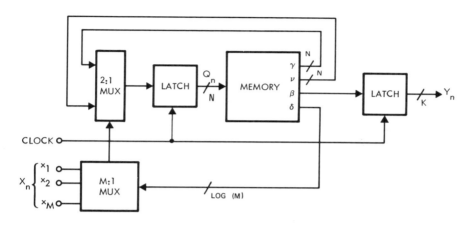

Figure 3-19. Multiplexer Based Microprogrammed Control Unit.

The subroutining goal in controllers is to permit frequently used blocks of microprograms to be stored once and "called up" as required for other routines. For example, several arithmetic routines in a signal processor may be such that a single output subroutine may be shared by all of them at a significant savings in microprogram size.

To implement subroutine operation it is necessary to provide a return address. The main program "jumps" to the first line of the subroutine and, at the same time, places the return address in a special register. The subroutine ends with a jump to the return address.

The multiplexer based controller of Figure 3-19 can be modified to provide single level subroutine operation, as shown on Figure 3-20. The modification involves replacing the 2:1 next address multiplexer with a 3:1 multiplexer which is used to select either γ or ν, the "conventional" next addresses, if the subroutine return bit is false, or the latched address (the subroutine return address) if the subroutine bit is true. The latch which stores the return address is not clocked while a subroutine is being executed. Replacing the subroutine return address latch with a last-in-first-out stack allows the implementation of multiple levels of nested subroutines.

Up to this point, the primary concern has been to reduce the size of the next state table of the Moore machine without unduly limiting the generality of the next state selection process. In addition to the obvious merging of microprogram

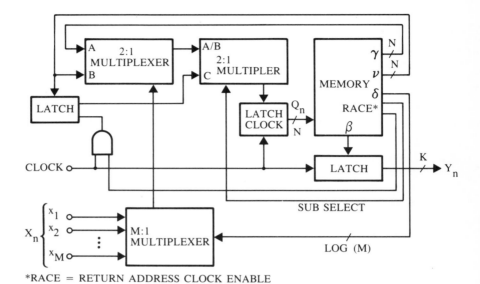

Figure 3-20. Multiplexer Based Microprogrammed Control Unit with Single Level Subrouting

rows or columns [3-11], it is appropriate to consider two possible techniques to reduce the microprogram size: encoding the output control lines [3-12], and employing Programmable Logic Arrays (PLAs) to generate the jump address. Output encoding requires selecting a set of control lines so that, at most, one is active at any time. For each machine state, only the number of the active output is stored in the microprogram. The control outputs are generated by an integrated circuit decoder at the controller output. Suppose there are 8 to 15 outputs; only 4 bits are required to encode the number of the active line. If none of the lines are active for some states (i.e., a no-op instruction), a dummy line must be provided which is activated when none of the "normal" controller outputs is active. This can be applied to the basic Moore machine output function or to any of the new implementations.

Using PLA or Read Only Memories (ROMs) to generate the jump addresses (i.e., ν and γ) involves encoding the addresses prior to storing the microprogram and then decoding them with a PLA or ROM. If, for example, there are 32 jump addresses used in a 1024 word microprogram, only 5 bits are required to encode the addresses as compared with 10 bits for the direct approach. The decoding requires a PLA or ROM with 5 inputs and 10 outputs. Practical experience in-

VLSI Architecture

dicates that jump address encoding is most useful with the presettable counter-based controller, since in the multiplexer based controllers the next address patterns are likely to be ill conditioned. A PLA is basically a "sparse" ROM, which is applicable to situations where only a small fraction of the possible binary input patterns are of interest.

The basic design of a controller for a small printer is described to illustrate the design process. A Seiko EP-101 printer is used in a status monitoring system because of its low cost, two color capability, and ready availability [3-13]. It is a 21-column printer with a hammer at each printing position, which is activated when the desired character is located on the drum beneath the paper.

The relationship between the status monitoring system, the printer, and the printer controller is shown in Figure 3-21. The status monitoring system generates Data Ready and Red/Black commands while the controller generates a reset data ready command and a four bit BCD code which gives the drum position. Once a printing cycle begins, the status monitoring system uses the drum position code to determine when to activate the printer hammer solenoids. The controller accepts timing pulses (one indicates when the drum has rotated to the next character position while the other is a frame synchronization pulse) from the printer and controls the motor and paper and ribbon advance solenoids.

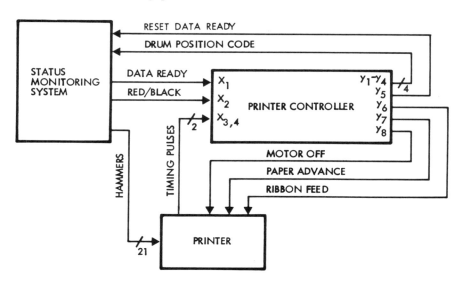

Figure 3-21. Printer Controller Interfaces

The controller state diagram is shown in Figure 3-22. Starting in state 0 (motor is off) when the data ready line is activated, the machine sequence through states 1, 2, and 3. This serves as a start-up delay allowing two times for the motor to reach operating speed. At state 3, a printer test circuit (which verifies the existence of the timing pulses) is interrogated. If the printer is operating correctly, the printing cycle begins; otherwise, the controller returns to 0 and tries again. States 4 through 22 are used to locate the printer frame synchronization pulse, which indicates that the drum is in the 0 position (i.e., the character 0 is under the hammers). Once the frame synchronization pulse occurs, the controller determines if the printing is to be red or black. If the printing is black, the drum position code may be sent to the status monitoring system. If the data is to be printed in red, the ribbon feed is activated prior to printing which latches the ribbon plate

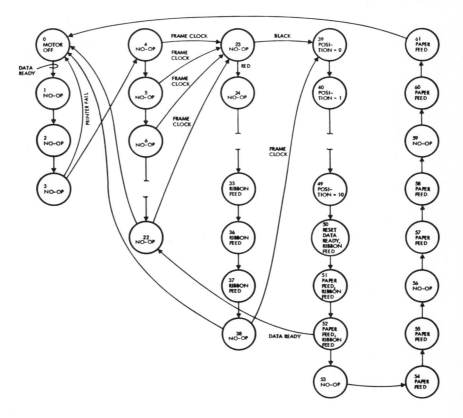

Figure 3-22. Printer Controller State Diagram

in the red position. This is cleared by a paper feed after printing. When the ribbon feed is completed, the frame synchronization pulse is located again and printing occurs as for black. After printing, the data-ready flip-flop is reset and a ribbon and paper feed are performed. If the data-ready flip-flop is set during the paper and ribbon-feed operation, more data is ready and the controller reverts to state 22 to print it without additional spacing (or the start-up delay); otherwise, three paper advances are performed so the newly printed line will be visible.

The multiplexer based controller of Figure 3-19 was used resulting in the final implementation shown in Figure 3-23. A 64 × 22 bit microprogram is required with a microprogram word format that consists of two next addresses (each 6 bits wide), 8 output lines, and 2 bits to identify the input bit which selects between the next addresses. The simplicity of this controller is apparent from the implementation complexity: Only six commercially available integrated circuits are required (two quad 2:1 multiplexers, one dual 4:1 multiplexer, and three 64 by 8 ROM). Thus a very simple design was produced with adequate flexibility for the specific application. This approach is equally applicable to most special purpose processing control requirements.

A range of implementations for specialized microprogrammed controllers has been presented. Although the early concepts of microprogramming were developed to realize general purpose processors, they have been modified to match the controller requirements of specialized processing systems. A basic Moore machine is implemented with a restricted state selection capability. This approach is suitable for many applications which do not require complex branching logic. It can be effectively realized with available MSI devices.

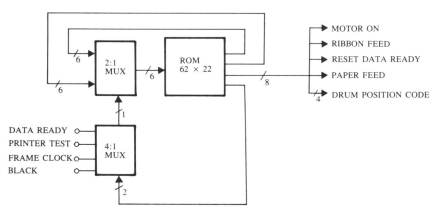

Figure 3-23. Printer Control Circuit

3.5 References

The Mead-Conway book [3-14] is in many ways responsible for the current interest in VLSI. By developing an introductory text suitable for University courses, it helped make VLSI accessable to a wide audience. More detail on the interaction between algorithm design and VLSI design is provided in [3-15].

3-1   Geroge H. Heilmeier, "Needed: A 'Miracle Slice' for VLSI Fabrication," **IEEE Spectrum,** March 1979, pp. 45-47.

3-2   Arthur L. Robinson, "Are VLSI Microcircuits Too Hard to Design?," **Science,** Vol. 209, July 11, 1980, pp. 258-262.

3-3   Gordon Moore, "VLSI: Some Fundamental Challenges," **IEEE Spectrum,** April 1979, pp. 30-37.

3-4   Jan Grinberg, Graham R. Nudd, and R. David Etchells, "A Cellular VLSI Architecture," **Computer,** January 1984, pp. 69-81.

3-5   Joseph Y. Lee, et al., "A 8 by 8b Parallel Multiplier in Submicron Technology," **International Solid-State Circuits Conference Digest of Technical Papers,** 1985, pp. 84 85, and 312.

3-6   Earl E. Swartzlander, Jr., "Microprogrammed Control for Specialized Processors," **IEEE Transactions on Computers,** Vol. C-28, 1979, pp. 930-934.

3-7   E.F. Moore, "Gedanken-Experiments on Sequential Machines," **Automata Studies,** Princeton: Princeton University Press, 1956.

3-8   M.V. Wilkes, "The Best Way to Design an Automated Calculating Machine," **Report of the Manchester University Computer Inaugural Conference,** Electrical Engineering Dept., Manchester University, Manchester, England, 1971, pp. 16-18, Reprinted in **Computer Design Development: Principal Papers,** ed. by E.E. Swartzlander, Jr., Rochelle Park, NJ: Hayden Book Co., 1976.

3-9   S.H. Fuller, V.R. Lesser, C.G. Bell, and C.H. Kaman, "The Effects of Emerging Technology and Emulation Requirements on Microprogramming," **IEEE Transactions on Computers,** Vol. C-25, 1976, pp. 1000-1009.

3-10 S. Dasgupta, "The Organization of Microprogram Stores," **Computing Surveys,** Vol. 11, 1979, pp. 39-65.

3-11 K. Hwang, "Fault-Tolerant Microprogrammed Digital Controller Design," **IEEE Transactions on Industrial Electronics and Control Instrumentation,** Vol. IECI-23, 1976, pp. 200-206.

3-12 A.K. Agrawala and T.G. Rauscher, **Foundation of Microprogramming,** New York: Academic Press, Inc., 1976, p. 62.

3-13 Earl E. Swartzlander, Jr., "Microprogrammed Sequential Machines," **New Components and Subsystems for Digital Design,** Santa Monica, CA: Technology Service Coroporation, 1975, pp. 111-114.

3-14 Carver Mead and Lynn Conway, **Introduction to VLSI Systems,** Reading, MA: Addison-Wesley, 1980.

3-15 Jeffrey D. Ullman, **Computational Aspects of VLSI,** Rockville, MD: Computer Science Press, 1984.

CHAPTER 4.
SIGNAL PROCESSING

An important application of VLSI is in the implementation of digital signal processing systems. This chapter examines signal processing algorithms, ranges of requirements, and architectures. It concludes with a description of the development of a sonar adaptive equalizer, an example of a typical signal processing system.

4.1 Signal Processing Algorithms

Digital signal processing algorithms include time domain and transform algorithms. Time domain processing includes digital filtering, correlation, digital encoding, and signal averaging. Transform processing includes various Fourier transform alorithms as well as transforms based on other basis functions.

There are two types of digital filters, recursive filters and FIR filters. Both are described briefly here. More information on filter history, design, and application can be found in [4-1].

Digital Filters

Recursive filters are generally used to realize infinite impulse response filters. One of the simplest examples of a recursive filter is the simple lowpass filter described by the equation:

$$Y(t) = (1-A) \, Y(t-1) + A \, X(t)$$

At any time t, the output Y(t) is a function of the present input X(t) and the previous output Y(t-1) where A is a convergence factor. Such filters are called recursive filters since the present output depends in part on the previous output. For finite precision arithmetic, a recursive filter output approaches the ideal output until the difference between the output and input is smaller than the system quantization. At this point the output ceases to change (although the value may differ from the correct final value). Recursive filters also may exhibit instability unless high precision arithmetic is implemented. Most applications require 16 to 24-bit arithmetic, although there is a strong tendency to consider floating point arithmetic for critical applications. Recursive filters are efficient in that they use only a small amount of digital hardware (multipliers, adders, and latches), but may be difficult to design and may suffer from instability unless high precision arithmetic is employed.

Finite Impulse Response (FIR) filters realize a finite impulse response and have important advantages over recursive filters such as phase linearity and modest arithmetic accuracy requirements. A FIR filter realizes an equation of the form

$$Y(t) = \sum_{i=0}^{n-1} X(t-i) K_i$$

where K is the filter kernel. Since the output of this filter does not depend on inputs occurring before the most recent n inputs; the filter it realizes has a finite impulse response. Generally, FIR filters do not require the high coefficient accuracy needed for recursive filters. As a result, arithmetic accuracy requirements are less severe. Chapter 5 presents a case study of FIR filter implementation using a variety of technologies and approaches.

Correlation

Correlation is widely used in signal processing (e.g., radar Barker code detectors, electronic warfare presorting, communication matched filters, terminal guidance of missiles, and linear predictive coding (LPC) voice bandwidth reduction systems). Basically, digital correlation involves computing a discrete approximation to the correlation function $R_{xy}(\tau)$

$$R_{xy}(\tau) = 1/T \int_0^T X(t)\, Y(t+\tau)\, dt$$

where X and Y are two continuous functions and τ is the displacement between X and Y. In sampled data digital correlation, the integral is replaced by summation

$$R_{xy}(\tau) = 1/K \sum_{k=0}^{K-1} X(k)\, Y(k+\tau)$$

where X and Y are digital waveforms. In many communication applications, X represents a known pseudonoise (PN) sequence and Y is the output of a receiver. The correlation of a PN sequence with a shifted version of itself is negligible unless $\tau = 0$. Thus correlating the receiver output with the known PN sequence can be used to synchronize the receiver with an independent transmitter. It also can be used in radar detectors. Here the PN sequence is a Barker word. In both of these cases, binary PN sequences are used; X is a fixed binary pattern while Y is a pattern obtained by thresholding the receiver output which produces a binary pattern.

Signal Averaging

Signal averaging algorithms include integration, averaging, standard deviation computation, noise spectrum estimation, etc. In general, at the output of a signal processing system, the user receives a set of measurements which have a significant probability of error. By using various averaging techniques, repeated sets of measurements can be combined to produce more reliable estimates. Generally, the SNR of each measurement improves as the square root of the number of measurements averaged. This is because the signal increases linearly while the noise (assumed noncoherent) increases as the square root of the number of measurements.

For simple integration, an adder and a delay line suffice. More advanced averaging algorithms, as used for constant false alarm rate and noise spectrum equalization, are often implemented with a specialized vector processor. The vector processor must be capable of summing arbitrarily weighted subsets of the input data stream (say 31 adjacent elements) and comparing a fraction of the weighted value with the central vector element. If the central vector element is large, it is assumed to be a bonafied signal. Various error metrics are used in the summation process (e.g., absolute value, summed squares, and Chebyshev or maximum). Signal averaging algorithms tend to vary widely between disciplines and are often implemented with programmable post-processors to facilitate program mode changes.

Transform Processing

In transform processing there are a few basic approaches that have been subject to continual refinement but have not changed drastically in nearly two decades.

The Fourier transform dominates the field of transform processing. The Discrete Fourier Transform (DFT) computes the spectrum of N digital data. The spectrum consists of the N components, F(0), F(1), . . ., F(N-1) where each is the correlation of the input data with complex sinusoids of frequencies from 0 to N−1. The kth spectral value is given by:

$$F(k) = \sum_{j=1}^{N} f_j \left(\cos \frac{2\pi jk}{N} + i\sin \frac{2\pi jk}{N} \right)$$

where: $i = \sqrt{-1}$. Thus computing one component of the spectrum requires N complex multiplications and N−1 complex additions. Since the complete spectrum includes N components, computing the complete spectrum using the DFT algorithm requires N^2 complex multiplications and approximately N^2 complex additions. Obviously as N becomes large the computation becomes difficult to implement, but it has a simple computational structure. An iteration of the algorithm proceeds as follows: the jth cosine/sine values are accessed, they are multiplied by the jth input data, and the complex product is added to the contents of an accumulator. This process is repeated N times to compute each spectral value. The algorithm is suitable for N level parallelism where N separate complex multiplier/accumulators are used, each evaluating a different component of the spectrum. Such parallelism is not possible for many data processing algorithms involving data dependent operation sequences, iterative computations, and interaction between data channels, but it is easily implemented for the DFT algorithm.

In 1965, the Fast Fourier Transform (FFT) algorithm was published [4-2]. Later it was realized that fast algorithms for the calculation of the Fourier transform have been independently discovered by Gauss [4-3] and many others, especially

Signal Processing

in the last two decades [4-4]. The basic notion is to factor the DFT computation so that it is implemented as a sequence of recursive operations called butterflys. The radix 2 butterfly operation consists of a complex multiplication and a sum and difference operation as defined by:

$$f'_i = f_i + f_j W_N^k$$

$$f'_j = f_i - f_j W_N^k$$

Where: W_N^k is a complex rotation coefficient defined by $W_N^k = \cos(2\pi k/N) + \sqrt{-1} \sin(2\pi k/N)$.

The butterfly operation is diagrammed in Figure 4-1. It is called a butterfly because its signal flow graph resembles a butterfly with its wings outstretched.

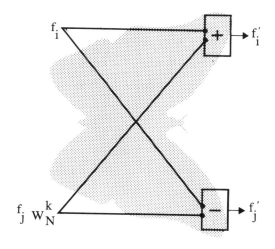

Figure 4-1. Radix 2 FFT Butterfly Operation

The basic operation of one form of the FFT algorithm is as follows. Consider an N point input sequence $f_0, \ldots, f_{N-1}$. A pair of adjacent data, starting with f_0 and f_1, are accessed, the second element of the pair, f_1, is multiplied by a cosine/sine term. Then the sum and difference of that product with the other input number, f_0, is formed. Those two results, f_0' and f_1' are placed back into memory, the next two values in memory, f_2 and f_3 are accessed and the process repeated until the entire sequence of N points has been processed as pairs of adjacent data. Then the process is repeated upon pairs of the results of the first pass that are separated by one, i.e., f_0' and f_2', f_1' and f_3', etc. The process is repeated upon pairs of intermediate results with successively doubled separation until the data has been processed with a spacing of N/2. There are log N passes each involving N/2 butterflies. With the FFT algorithm the total computation is N/2 log N butterflies, each consisting of a complex multiplication and two complex additions.

Table 4-1 compares the number of operations for the DFT and the FFT. For all values of N the FFT requires fewer operations, but the advantage is more significant for large values of N. For example, for N = 1024, the DFT requires approximately 2,000,000 complex arithmetic operations while the FFT requires only 15,000 complex arithmetic operations. This advantage is obtained at a price however, since the FFT requires memory to store intermediate results and requires a more complex structure to generate the addresses and to control the processor.

	DFT			FFT		
N	Complex Additions	Complex Multiplications	Total Operations	Complex Additions	Complex Multiplications	Total Operations
4	16	16	32	8	4	12
16	256	256	512	64	32	96
64	4K	4K	8K	384	192	576
256	64K	64K	128K	2K	1K	3K
1024	1M	1M	2M	10K	5K	15K
4096	16M	16M	32M	48K	24K	72K

Table 4-1. Comparison of DFT and FFT

4.2 Performance Requirements

Transform processing performance requirements are typical of signal processing. The three parameters of interest are speed, dynamic range, and frequency resolution. Table 4-2 shows the range of functional requirements that arise in transform processing.

The speed is measured by the input data rate in samples per second. It varies from 10,000 samples per second or lower for sonar applications to communications and radar applications that require 10,000,000 or more samples per second, with some radar applications that approach 100,000,000 samples per second.

Signal Processing

System Parameter	Processor Parameter	Size		Application
Speed	Input Data Rate (Samples/sec)	10,000		Sonar
		1,000,000		Radar, Communications
		10,000,000		Radar
Dynamic Range	Word Size (Bits)	Fixed Point	4 to 8	Image Compression
			12	Radar, Communications
			16	Radar, Sonar
		Floating Point	16	Communications
			32	Digital Voice
Frequency Resolution	Memory Size (Words)	32 — 64		Image Compression
		256 — 1024		Radar
		4096 — or more		Sonar

Table 4-2. Spectrum Analysis Requirements

The second parameter is dynamic range. This results in requirements for larger arithmetic word size. There are cases in image compression and some radar applications where 12-bit (or even lower) precision arithmetic is acceptable, but most systems at the present time require at least 16-bit precision. There is a strong trend toward floating point arithmetic with word sizes of at least 22 bits.

The 22-bit format, with a 16-bit two's complement fraction and a 6-bit two's complement exponent, is a reasonable compromise among performance, speed, and size. Although single chip 32-bit floating point devices are commercially available, for a given technology, the 22-bit format will always produce chips that are simpler and, as a result, more producable and faster [4-5] with adequate dynamic range and precision for most applications. Floating point wordsizes of 32 bits and larger are useful in scientific computation when inverting matrices, evaluating eigenvectors, etc., but these operations are usually performed at much lower rates than those required for signal processing where input data are often limited in precision to eight bits or less.

The chief advantage of floating point arithmetic is increased dynamic range. As shown on Table 4-3, 22-bit floating point arithmetic provides 96 dB of precision (i.e., equivalent to 16-bit fixed point arithmetic) over a dynamic range of 480 dB. Although this dynamic range is less than that of 32-bit floating point arithmetic, it is more than adequate for most high speed signal processing applications. The 16-bit fixed point format has a dynamic range of $16 \times 20 \ln 2 = 96$ dB; the 22-bit floating point format provides $(2^6 + 16) \times 20\ln 2 = 480$ dB. With proper normalization, 16-bit precision is available over a dynamic range of $2^6 \times 6 = 384$ dB. In contrast, 22-bit fixed point would offer 16-bit precision over only a 6-bit (36 dB) range.

Arithmetic System	Dynamic Range	Precision
12-Bit Fixed Point	72 dB	72 dB
16-Bit Fixed Point	96 dB	96 dB
22-Bit Fixed Point	132 dB	132 dB
22-Bit Floating Point	480 dB	96 dB
32-Bit Floating Point	1686 dB	144 dB

Table 4-3. Arithmetic Comparison

The final parameter for use in categorizing spectrum analysis systems is the frequency resolution which manifests itself in the amount of internal memory that is required. The ratio of the frequency range divided by the resolution is the minimum transform length. It ranges from 4,096 to 16,384 for sonar applications down to 32 to 64 for image compression.

By way of an example, the required performance to achieve an example processor is estimated for the functional performance and gate-rate metrics of Section 2.1. For the example, assume a radar processor that receives 10^6 samples per second, uses 16-bit fixed point arithmetic, and performs 1024 point transforms. For each transform, 1024 data are processed by computing 5,120 radix 2 butterflys, an increase by a factor of five relative to the input rate. Thus, to process 10^6 samples per second requires 5×10^6 butterflys per second. A butterfly is implemented with four real multipliers and six real adders according to the general structure of Figure 4-1. Since 10^7 multipliers per second are achieved with commercial 16-bit multipliers as indicated in Section 3.2, it should be possible to implement this FFT processor with 25 to 50 commercial circuits (10 for the arithmetic, 10 for interfaces, and 5 miscellaneous control circuits).

Since a butterfly requires four mulitpliers and six adders and since the gate complexity of an n bit multiplier is approximately $20n^2$ gates and the complexity of an n bit adder is approximately $20n$, the butterfly gate complexity is approximately 22,000 gates. If pipeline registers are used between arithmetic operation, the multipliers and adders must be fast enough to complete a multiplication or addition in 2×10^{-7} seconds. Multiplying the reciprocal of the multiplier delay times the required number of gates gives a total gate-rate of approximately 10^{11}. Since the commercial multiplier of Section 3.2 achieves a gate-rate of 5×10^{10}, this analysis suggests that on the order of one to ten circuits will be required to implement the example FFT processor, a figure that is in excellent agreement with the estimate developed with the functional performance metric.

4.3 Signal Processing Architectures

This section examines some signal processing architectures. First, the two classical general purpose computer architectures are examined to provide a good understanding of the important issues. Both the von Neumann and Harvard architectures for general purpose computers have been widely applied to signal processing especially for moderate speed sonar applications. This section also examines programmable signal processors and approaches for developing custom signal processors. The modular approach assumes board sized processing modules which will eventually be replaced by single integrated circuits as the chip complexity increases.

von Neumann Architecture

The von Neumann architecture consists of a memory, an arithmetic unit, and a control unit. As shown on Figure 4-2, the memory is connected to a bus which routes data and addresses to the memory from the aritmetic unit and the control unit. The bus also routes data and instructions from the memory to the arithmetic unit and the control unit. Such a system operates by fetching an instruction from

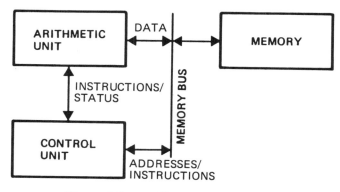

Figure 4-2. von Neumann Architecture

the memory to the control unit, decoding the instruction, fetching data from the memory to the arithmetic unit, performing the instruction on the data, and returning the results of the arithmetic operation to memory. Another instruction is fetched and the process is repeated. If the average instruction requires accessing one value of data from memory for operation in the arithmetic unit, then the memory and the memory bus must be twice as fast as either the arithmetic or control unit. This is an attractive architecture from the standpoint that it uses only a single memory unit which stores both the instruction codes and data [4-6], but it suffers from the potential bottleneck of the memory bus which needs to be faster than the remainder of the system. Presumably these elements will be implemented with a more advanced technology than that used with either the arithmetic or control unit. This architecture was developed in the context of the very early computers when memory was the pacing element in computer design. In that context, this is an extremely clever architecture because this architecture uses only one memory and shares it amongst all the elements of the system.

Harvard Architecture

In constrast to the von Neumann architecture, the Harvard architecture uses two separate memories as shown in Figure 4-3; a data memory which is coupled directly to the arithmetic unit and an instruction memory which is coupled to the control unit. This eliminates the memory bottleneck, but it requires two separate memories. A potential advantage with current technology is that in signal processing the instruction sequence often remains fixed so that a ROM may be used for the instruction memory. A separate read/write RAM memory is used to sup-

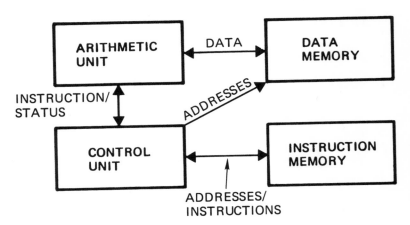

Figure 4-3. Harvard Architecture

port the data memory requirements for the arithmetic unit. Replacing a portion of the RAM in the von Neumann architecture with ROM, which is inherently simpler to implement, is advantageous. In comparing the von Neumann and Harvard architectures, the only real difference is the number of memories: the single memory von Neumann architecture is susceptible to bottlenecks, but minimizes the number of memory units that are required. The Harvard architecture separates the data stream from the instruction stream which reduces the speed required for the memory and the bus. The speed advantage of the Harvard architecture may be mitigated by the need to provide additional package pins for the multiple memory ports.

Programmable Signal Processor Architecture

One of the first widely used structures for signal processing was the Programmable Signal Processor (PSP), shown in Figure 4-4. PSP-based systems use a general purpose computer (to provide programmability and flexible interfaces) augmented with a specialized processor (to provide high speed arithmetic). During operation, the general purpose computer takes in data from the sensors and loads the PSP memory. It also places a sequence of instructions in the PSP memory at a prescribed set of locations. The PSP arithmetic unit then executes those instructions on the data at a high rate. The PSP has an arithmetic unit tailored to signal processing requirements allowing it to be fast and relatively simple. It sequences through the data in memory according to the prescribed instructions plac-

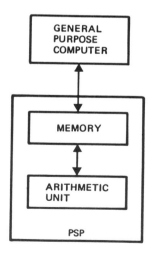

Figure 4-4. Programmable Signal Processors

ing the results in specified locations. The general purpose computer then accesses those results, provides appropriate post processing, and sends them on to their destination. The specialized processor is tailored specifically to signal processing. The general purpose computer serves as the controller for the system, implements the data input and output operations, generates the instruction sequence for the PSP, and it does most of the data dependent processing e.g., looking for peaks in the data and detecting narrow spectrum components, etc. The general purpose computer provides flexibility and facilitates interfacing the PSP to a variety of applications, but may be a bottleneck since all data passes through it. Some PSPs alleviate this problem by using multiple data channels to handle data interfacing. See [4-7] for a comparison of several early signal processing computers including the SPS-41, one of the first commercial PSPs.

Modular Signal Processor Architecture

The modular signal processor architecture is an approach that exploits the advantages of the Harvard architecture for signal processing systems [4-8]. This approach builds signal processing systems from a number of processing modules. The modules may be spectrum analysis modules, digital filters, input modules, bulk memories, etc. that are optimized for a single type of function tending to make them very efficient. For example, a digital filter module will be designed to implement one of a particular class of digital filters, say a Finite Impulse Response (FIR) filter. Those processing modules will have separate ports for data and control. As shown on Figure 4-5, the data is routed from module to module with dedicated connections as appropriate for the particular problem. The system is controlled via the control microprocessor which issues commands to the processing modules. The control microprocessor provides some flexibility since it controls the instruction sequence executed by the processing modules. This is

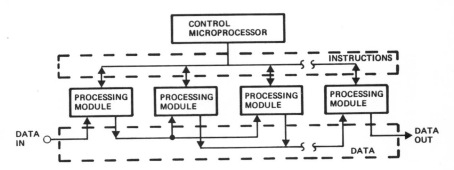

Figure 4-5. Modular Signal Processor

Signal Processing

an efficient architecture since the data routing from one processing module to another is direct (and application specific) and the processing modules are optimized for the specialized functions they perform. It requires a greater investment in nonrecurring (i.e., design) costs since a typical application will require several types of processing modules. At present, the modules are best realized as board level designs implemented with VLSI circuits as appropriate. In the future, as integration levels increase, it may be possible to create single chip processing modules.

The modular signal processor architecture is best understood by examining two spectrum analysis examples. The first implements the DFT computation. With the DFT, as a block of data enters the system it is multiplied by a particular set of cosine/sine values to compute one component of the spectrum. For each additional component of the spectrum a different set of cosine/sine values are used with the same data. As shown on Figure 4-6, an input buffer brings in data, fans it out in parallel to a number of DFT modules which send their results to an out-

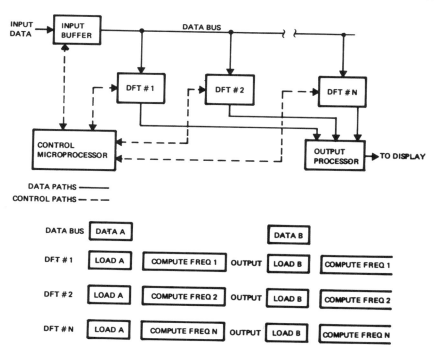

Figure 4-6. DFT Spectrum Analyzer

put processor. As shown in Chapters 6 and 7, the DFT modules range in complexity from a single VLSI circuit to tens or hundreds of circuits depending on the speed requirements and the available technology. The computation of the spectral components are performed independently in separate DFT modules.

A similar architecture can be used to perform spectrum analysis using the FFT algorithm. In this implementation the FFT is treated as a batch process where the FFT processor takes in a block of data, computes for some time (perhaps an order of magnitude longer than a DFT processor requires to compute one spectral component), and when it is done sends the complete spectrum to the output processor. Figure 4-7 shows how the same basic architecture used for the DFT spectrum analyzer is used to implement the FFT. The data is brought in through the input buffer, routed via the data bus, and loaded in sequence into the various FFT processors. The first block of data, data block A, is loaded into FFT processor number one, which starts computing the spectrum. Then the next block

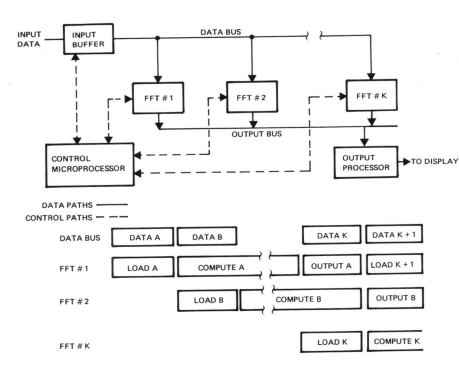

Figure 4-7. FFT Spectrum Analyzer

Signal Processing

of data comes in. Since FFT processor number one is busy, the second block of data, data B, is loaded into processor two. As successive blocks of data arrive, each is loaded into the next available processor. If enough processors are available a stable situation develops where each processor is finished before it is needed for an incoming data block.

Several of the required processing modules may be effectively implemented with generic architectures. As an example, consider the use of the multiport memory (introduced in Section 3.4). As shown in Figure 4-8 the multiport memory is coupled directly to an arithmetic function with the results coupled back into the memory. In this architecture two operands are accessed from the memory, loaded into the arithmetic function, a computation is performed, and the result is loaded into the memory. This generic architecture is also applicable to FFT computation, where the radix 2 butterfly operation requires two operands and generates two results. Using the architecture of Figure 4-8, two data are accessed from memory, the butterfly computation is performed, and the two results are returned to the memory. Thus the memory needs two read ports and two write ports. For applications other than the FFT, where only one result is computed by the arithmetic function, only a single write port is needed. In these applications the other write port can be used for loading data into the memory while data from another portion of the memory is being processed. When processing is completed, the two portions of memory switch. The newly arrived data is processed while new data is loaded.

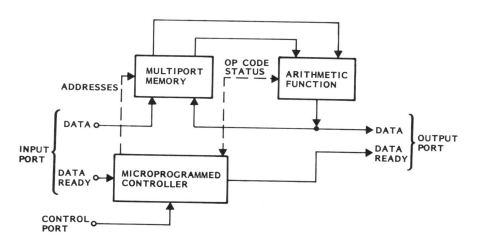

Figure 4-8. Signal Processor Architecture

The generic architecture can also accommodate multiple arithmetic units. Figure 4-9 shows an example using a multiplier accumulator and a bit slice microprocessor. If the multiport memory is faster than the arithmetic units, a ping pong computation pattern can be used. In this mode, a pair of data from the memory are loaded into one arithmetic unit. On the next cycle, another pair of data from the memory are loaded into the other arithmetic unit. On successive cycles, results from each arithmetic unit are stored in the memory as another pair of operands are fetched. This process can continue indefinitely with one fast multiport memory shared between multiple arithmetic units.

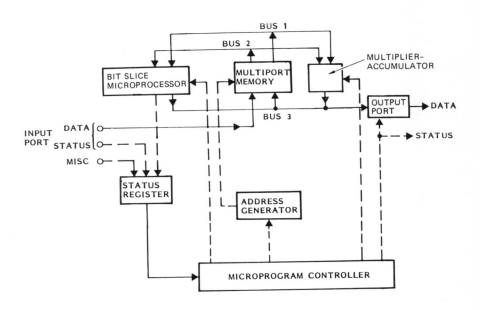

Figure 4-9. High Performance Signal Processor Architecture

The modular signal processor architecture assumes fixed data routing between signal processing modules. In the future as networking (introducing in Chapter 8) matures, it will be appropriate to use restructurable networks to provide the data interconnections between modules.

Signal Processing

4.4 Adaptive Sonar Equalization

The concept of adaptive sonar equalization is illustrated on Figure 4-10. Raw data sensed with a hydrophone is shown on Figure 4-10A, which is a plot of acoustic pressure versus time. The spectrum of these data are shown in 4-10B, a plot of spectrum intensity versus frequency. The spectrum consists of a broad and relatively smooth continuum which has superimposed upon it a variety of sharp features (i.e., narrow spikes, hills, valleys, etc.). This continuum indicates the approximate acoustic conductivity of the ocean at each frequency: higher values indicate that the water transmits sound at those frequencies with less attenuation. The sharp features correspond to acoustic signals that may have been generated by boats, ships, large fish, etc. The amplitude of these spikes is an indicator of the signal intensity, provided that corrections are made to account for the frequency dependent acoustic conductivity. Such correction involves developing an estimate of the acoustic conductivity by smoothing the noise continuum as shown in Figure 4-10C. This estimate of the acoustic conductivity is divided into the raw spectrum to produce the equalized spectrum shown in 4-10D. Using the equalized spectrum, an analyst can estimate the distance to certain noise sources because the amount that the signal decreased while traveling through the water can be estimated from the received signal intensity.

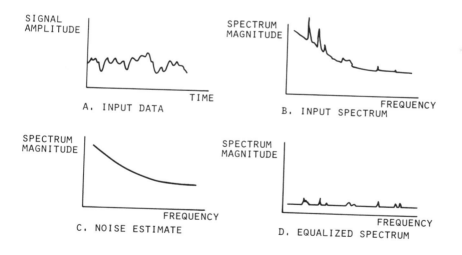

Figure 4-10. Adaptive Sonar Equalization

The adaptive equalizer is implemented as shown in Figure 4-11. The spectrum is computed in the spectrum analysis unit, which windows the data in the time domain, performs an FFT, and computes the magnitude of the spectrum. The magnitude is integrated to estimate the noise level. That noise estimate is divided into the incoming spectra to generate the equalized output. With the equalized output the received signals are normalized to represent equal signal strengths independent of the frequency.

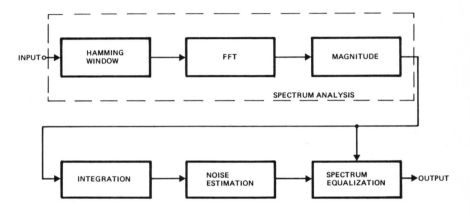

Figure 4-11. Sonar Equalization Algorithm

In this system most of the computation is being done within the FFT which represents 90% of the total computational load. The remaining aspects, data windowing, magnitude computation, and noise estimation all impose relatively low computational loads. This is fortunate since the noise estimation is a rather difficult process. It requires averaging several successive incoming spectra, editing to null out peaks, and averaging over adjacent frequencies to fill valleys, etc. It is highly data-dependent but is performed so infrequently that the FFT which is computed on every block of input data is the dominant computational load.

Figure 4-12 shows a design for an adaptive sonar equalizer. It uses three signal processing modules with direct interconnection and a microprocessor based controller. The input module consists of an A/D converter, some control logic, and a memory. Analog data is digitized and loaded into the memory until there is a complete block of data. Then the data is transferred to the transform processor. The transform processor consists of a parallel multiplier, a bit slice microprocessor,

Signal Processing

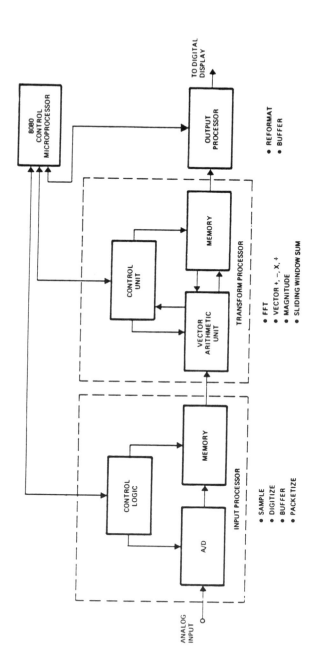

Figure 4-12. Sonar Equalization Implementation

and a working memory. The memory is coupled very closely to the arithmetic unit via a microprogrammed control unit. Finally there is an output processor that takes the resulting equalized spectra and stores up several different frames of the output data so that they can be displayed on a CRT or output to a plotter. The whole system operates under the control of an a microprocessor which generates the stream of control commands and synchronizes the system. Most of the computation is performed by the dedicated signal processing elements; in this case, the input processor, the transform processor, and the output processor. Those three elements are tailored for the restricted set of operations they perform so that they are very efficient. The microprocessor provides flexibility but since it is not in line with the data computations its speed is not critical.

The advantage gained by separating the control from the computation via the use of the microprocessor and the specialized signal processing modules is illustrated by an incident that occurred as the project was nearing completion. It was discovered that the wrong noise estimation algorithm had been implemented. The microprocessor was reprogrammed to reorder the sequence of operations that were being performed in the transform processor so that the correct noise estimation algorithm could be demonstrated two days later.

4.5 References

There has been a progression of excellent texts in digital signal processing. The book by Gold and Rader [4-9] served as the introductory text through the mid 1970s. As a text it has been supplemented by the much more detailed information contained in [4-10] and [4-11]. Hardware implementation is emphasized in [4-12], [4-13], and [4-14]. The sequence of selected reprints contained in [4-15] and [4-16] provide an excellent overview of design, analysis, and implementation developments in digital signal processing.

4-1   R.W. Hamming, **Digital Filters**, Englewood Cliffs: Prentice-Hall, 1977.

4-2   James W. Cooley and John W. Tukey, "An Algorithm for the Machine Calculation of Complex Fourier Series," **Mathematics of Computation**, Vol. 19, 1965, pp. 297-301.

4-3   Michael T. Heideman, Don H. Johnson, and C. Sidney Burrus, "Gauss and the History of the Fast Fourier Transform," **IEEE ASSP Magazine**, Vol. 1, No. 4, October 1984, pp. 14-21.

4-4 H.J. Nussbaumer, **Fast Fourier Transform and Convolution Algorithms**, New York: Springer-Verlag, 1982.

4-5 John A. Eldon and Craig Robertson, "A Floating Point Format for Signal Processing," **Proceedings IEEE International Conference on Acoustics, Speech, and Signal Processing**, 1982, pp. 717-720.

4-6 Arthur W. Burks, Herman H. Goldstine, and John von Neumann, **Preliminary Discussion of the Logical Design of an Electronic Computing Instrument**, Princeton: Institute for Advanced Study, 1946. Reprinted in A.H. Taub, ed., **John von Neumann: Collected Works**, Vol. 5, New York: Pergamon Press, 1961, pp. 34-79.

4-7 Jonathan Allen, "Computer Architecture of Signal Processing," **Proceedings of the IEEE**, Vol. 63, 1975, pp. 624-633.

4-8 Earl E. Swartzlander, Jr., "Software and Firmware for Distributed Signal Processing," **Proceedings of the IEEE Micro Processors in Military and Industrial Systems Workshop**, 1980, pp. 47-54

4-9 Bernard Gold and Charles M. Rader, **Digital Processing of Signals**, New York: McGraw-Hill, 1969.

4-10 Lawrence R. Rabiner and Bernard Gold, **Theory and Application of Digital Signal Processing**, Englewood Cliffs: Prentice-Hall, 1975.

4-11 A.V. Oppenheim and R.W. Schafer, **Digital Signal Processing,** Englewood Cliffs: Prentice-Hall, 1975.

4-12 A. Peled and B. Liu, **Digital Signal Processing**, New York, John Wiley & Sons, 1976.

4-13 B. A. Bowen and W. R. Brown, **VLSI Systems Design for Digital Signal Processing**, Englewood Cliffs: Prentice-Hall, 1982.

4-14 Andres C. Salazar, ed., **Digital Signal Computers and Processors**, New York: IEEE Press, 1977.

4-15 Lawrence R. Rabiner and Charles M. Rader, **Digital Signal Processing**, New York: IEEE Press, 1972.

4-16 A. V. Oppenheim, et al., eds. **Selected Papers in Digital Signal Processing, II**, New York: IEEE Press, 1976.

CHAPTER 5.
DIGITAL FILTER CASE STUDY

This chapter, based in part on [5-1], is a case study examining the implementation of a Finite Impulse Response (FIR) digital filter.

5.1 FIR Filter Requirements

As noted in Section 4.1, the output of a FIR filter Y(t) is the summation of the n most recent inputs X(t), X(t−1), ..., X(t−n+1) weighted by a filter kernel function, K_i.

$$Y(t) = \sum_{i=0}^{n-1} X(t-i)K_i \qquad (5\text{-}1)$$

The inputs are multiplied on a point-by-point basis by the kernel function to generate each filter output value. The filter computes an inner product between the input vector X and the filter kernel vector K. One possible architecture to implement the filter is shown in Figure 5-1. Data enters a series of shift registers. X(t) is at the input of the first shift register, X(t−1) is at the output of the first shift register, X(t−2) is at the output of the second stage, X(t−3) is at the output of the third stage, etc. Each of the n elements of the X vector is multiplied by the appropriate kernel value and the products are summed with an n input adder. This simple parallel approach uses hardware in proportion to the filter length,

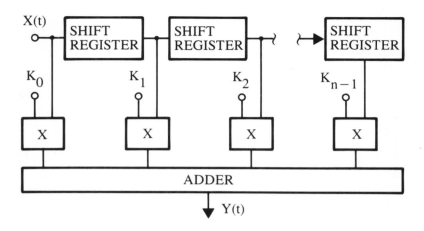

Figure 5-1. Finite Impulse Response (FIR) Filtering

i.e., an n point filter uses $n-1$ shift register stages, n multipliers, and an n input adder. Such a filter gives very precise control over the response, but it will be quite complex if n is large and a fully parallel implementation is used.

For many applications, the kernel function, K, is symmetric about its center point, so that the right side is a mirror image of the left side. Such symmetric kernel functions produce filters with minimum phase distortion. For the symmetric filter kernel case, the filter equation can be factored resulting in a summation from $i=0$ to $n/2$, (i.e, half as many terms) where each term is the sum of the two input values that are multiplied by the same kernel value. The process is shown on Figure 5-2: the long data shift register is folded in the middle, thus aligning input data that are equally distant from the centerpoint. Those data are added together, multiplied by the filter kernel value, and the products are summed. The advantage of this approach is a reduction in complexity since only half as many multipliers are required. The "extra" $n/2$ adders, before the multipliers, exactly offset the savings in complexity in the final adder which now has $n/2$ inputs instead of n as in the implementation in Figure 5-1. The net complexity is $n-1$ shift register stages, n adders, and $n/2$ multipliers.

Although the details are beyond the scope of this section, it is possible to further simplify the implementation of FIR filters by exploiting zeros or binary fractions (i.e., ½, ¼, etc.) in the kernel function. Such kernel values allow the corresponding multipliers to be eliminated or replaced with simple shift networks.

Digital Filter Case Study

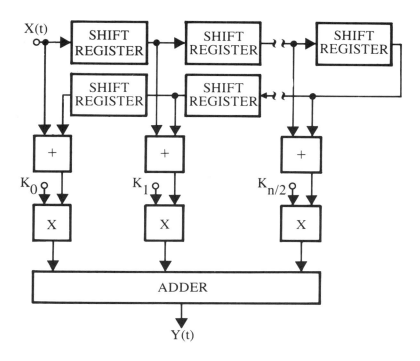

Figure 5-2. Symmetric FIR Filter

In comparing Figures 5-1 and 5-2, it is clear that the lower portion of the structure (i.e., the multipliers and the final adder) is common to both filters. This structure computes an inner product between the incoming data and a reference kernel function.

5.2 Filter Architecture

The FIR filter is implemented with a structure consisting of several multipliers whose outputs are summed with an adder. This is a direct implementation of the structure that is common to both asymmetric and symmetric kernel filters. The concern is how to form a number of parallel products and sum them to produce a single result. One architectural approach is shown on Figure 5-3. Inputs consisting of k pairs of data are multiplied and the products summed. An extra adder is included at the output for expansion. The final adder output can be latched and fed back into the adder creating an accumulator. The first k data pairs are

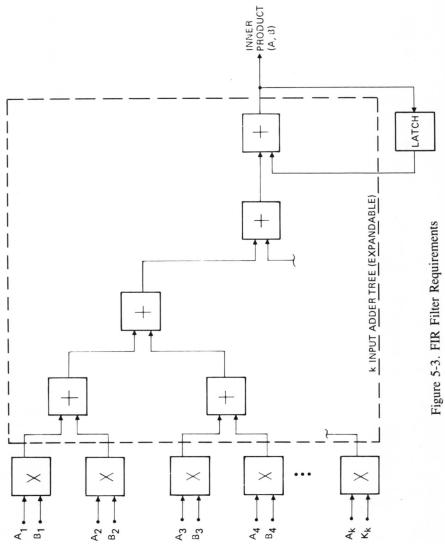

Figure 5-3. FIR Filter Requirements

Digital Filter Case Study

multiplied and summed to get a partial result which is held in the latch. Then the next k data pairs are processed to get a second result that is added to the previous partial result (from the latch output) and the new result is held in the latch, etc. This is an attractive structure for comparing different technologies since improvements in either speed or circuit gate count produce better systems performance (as measured by the amount of computation.)

For a specific application with filter length n, the required speed is n/k times the input data rate. Thus, FIR filters implemented with a limited density technology where k is small must operate at a speed-up relative to the input data rate, while a higher density implementation with a larger value of k can operate at lower rates.

The digital filter is defined by the following equations:

$$\left. \begin{array}{l} Y(1) = X(0)K_1 + X(-1)K_2 + X(-2)K_3 + X(-3)K_4 + \ldots \\ Y(2) = X(1)K_1 + X(0)K_2 + X(-1)K_3 + X(-2)K_4 + \ldots \\ Y(3) = X(2)K_1 + X(1)K_2 + X(0)K_3 + X(-1)K_4 + \ldots \\ Y(4) = X(3)K_1 + X(2)K_2 + X(1)K_3 + X(0)K_4 + \ldots \end{array} \right\} \quad (5\text{-}2)$$

Most implementations are "output driven" where the first output $Y(1)$ is computed based on inputs $X(0)$, $X(-1)$, ..., $X(1-n)$, then output $Y(2)$ is computed, etc.

5.3 FIR Filter Implementation Options

In this section four candidate FIR filter implementations are examined in detail. The four options are summarized in Table 5-1. The first approach is based on a commercial multiplier accumulator (as described in Section 3.1) implemented with a 2 micron design rule TTL compatable bipolar technology. The second approach uses merged arithmetic (an extension of one of the techniques used to construct very fast multipliers) which is implemented using an advanced (1 micron design rule) bipolar technology. The third approach uses unconventional arithmetic based on the use of logarithms. As is well known, logarithms are convenient for multiplication, but as shown in this section they can also be used to perform addition. This approach is realized with the one micron advanced bipolar technology used for the merged arithmetic approach. The fourth approach is the modular array, an approach based on an ultra fast silicon Emitter Coupled Logic (ECL) technology with limited logic density. Use of such a limited density technology requires development of a building block structure that is replicated to implement the FIR filter.

Approach	Technology	Gate (Rate)
Commercial Multiplier-Accumulator	2 micron Triple Diffused	10K Gates (7 MHz)
Merged Arithmetic	1 micron Triple Diffused	100K Gates (20 MHz)
Sign/Logarithm Convolver	1 micron Triple Diffused	100K Gates (20 MHz)
Modular Array	ECL	1K Gates (500 MHz)

Table 5-1. FIR Filter Options

Thus, the first approach uses commercial TTL, the second and the third use advanced bipolar technology, and the fourth uses very high speed custom ECL. In comparing the options, a functional performance metric as described in Section 2.1 is used. The figure of merit is the number of terms multiplied per second within the digital filter divided by the implementation complexity (i.e., the chip count). The goal is to select the implementation that provides a given level of performance with the fewest chips. Since cost, power consumption, size, and failure rate are, at least to a first order, proportional to chip count maximizing this metric maximizes the system performance by these other criteria as well.

Commercial Multiplier-Accumulator Implementation

The first approach is to use the commercial multiplier-accumulator. The design of this chip is a simple extension of the parallel multiplier mentioned in Section 3.3.

The structure is shown in Figure 5-4. It consists of a multiplier followed by an adder, which is followed by a register which feeds back to the adder. The figure also shows the control lines and some of the logic that is provided for multiplexing the inputs and outputs.

The FIR filter is implemented with the multiplier-accumulator by supplying appropriate pairs of data to compute a given filter output according to equation 5-1, i.e., $X(0)K_0$, $X(-1)K_1$, $X(-2)K_2$, etc. The multiplier-accumulator forms the product of each pair of terms and sums it partial filter output with the accumulator. When $Y(0)$ has been computed, the process is repeated for one later X value producing $Y(1)$, etc. This is equivalent to computing the system of equations (5-2) on a row at a time basis. The multiplier-accumulator is implemented

Digital Filter Case Study

with commercial technology which operates at an 8 MHz clock rate. Thus it is capable of computing 8×10^6 filter terms per second.

Merged Arithmetic Implementation

The second approach is based on the use of merged arithmetic. This involves a structure based on an extension of Dadda's approach for fast multiplication [5-3].

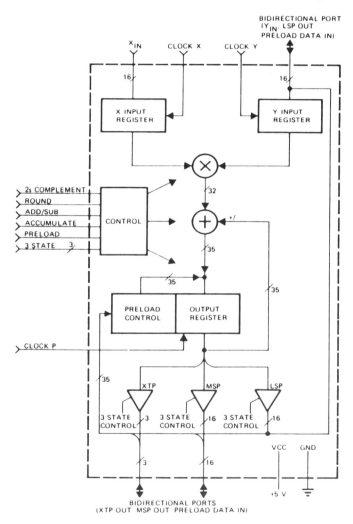

Figure 5-4. Commercial Multiplier Accumulator

Fast multiplication forms a product of two numbers in a number of delays that is proportional to the log of the number of bits. For a 16-bit fast multiplier it takes 5 delays. Since a conventional 16-bit multiplier would require 32 delays there can be a significant savings. The fast multiplier algorithm is directly extended to implement inner products. With merged arithmetic composite arithmetic functions are generated directly without forming distinct arithmetic operations (i.e., additions, subtractions, and multiplications) [5-3]. The approach will be described for positive numbers. It is easily modified to provide two's complement operation by inverting some of the bit products [5-4] and [5-5].

Computing an inner product of two k-element (each of M bits) vectors via merged arithmetic involves three distinct steps. First, the bit product matrix is generated with an array of kM^2 AND gates. In the second step, the matrix is reduced by counting the 1 bits in each column and performing carry processing to produce a two-row matrix. Finally, the two rows are summed in a carry-lookahead adder to generate the desired inner product.

The technique used to reduce the bit product matrix to a two-row intermediate matrix is illustrated for a two-term (each 8 bits) inner product (k = 2, M = 8) in Figure 5-5. At the top of the figure, the bit product matrix is shown as two parallelogram-shaped arrays of 8 × 8 dots (each dot represents one of the KM^2 bit products). The outputs of a full adder are indicated by two dots connected with a diagonal line, and the outputs of a half adder are indicated by two dots connected with a "crossed" line. Since a full adder accepts three bit values from a column and generates a sum bit in the same column and a carry bit in the next most significant column, the number of bits in a given column is reduced by two for each full adder, and the number of bits in the matrix is reduced by one. Dadda has developed an optimum strategy for minimizing the number of full adders required to implement a multiplier [5-2] which provides equally optimum results for merged inner product computation. Dadda's concept has been extended to take advantage of recent technology including fast high-capacity ROM [5-6].

The total delay for computation of the result of two k-term (each M bits) FIR filters is the sum of the delay of the three steps (i.e., the sum of the bit product gate delay, the adder network delay, and the carry-lookahead adder delay). The dominant delay is that of the adder network; the number of adders in the longest path is equal to the largest j such that

$$kM > d_j$$

where

$$d_1 = 2 \text{ and } d_j = [3d_{j-1}/2]$$

Digital Filter Case Study

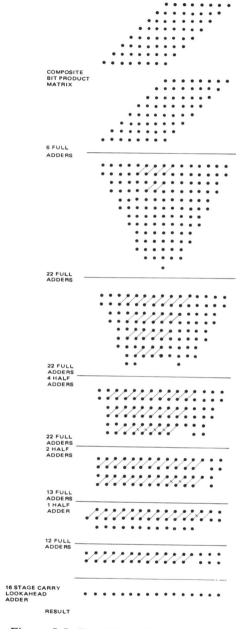

Figure 5-5. Two-Term Merged Multiplier Adder

Note that [X] denotes the largest integer less than X. An approximate value of j is given by

$$j \approx [2 \log (kM) - 2].$$

Thus, the total delay of a k-term multiplier-adder is

$$T_{mult} \approx T_{gate} + T_{add}[2 \log (kM) - 2] + T_{cla} (2M + \log k) \qquad (5\text{-}3)$$

where T_{gate} is the delay of a gate, T_{add} is the delay of an adder module, and $T_{cla(I)}$ is the delay of an I-bit carry-lookahead adder.

The number of full adders may be computed by noting that each full adder reduces the number of bits in the bit product matrix by one since a full adder has three inputs and two outputs. Since there are kM^2 inputs and $(4M + 2[\log(k - 1)])$ outputs, the number of full adders is bounded above by $(kM^2 - 4M)$. Empirically, a small number of half adders (approximately M) is generally required, so that the total adder count N_{add} is approximately

$$N_{add} = kM^2 - 3M.$$

An example of the merged arithmetic implementation of a four-term FIR filter computational element is shown in Figure 5-6. The element provides the functions of a conventional circuit consisting of four multipliers and an adder. The total hardware requirement for the merged arithmetic implementation is

128 two-input AND gates

104 adder modules (97 full adders and 7 half adders)

One 17-bit carry-lookahead adder.

In comparison to conventional direct methods, a merged implementation generates all of the bit products "en masse" using a single reduction network. This approach reduces the number of carry-lookahead adders from $2k - 1$ (k are used at the output of each multiplier and $k - 1$ are used as an adder tree) to one at a cost of increasing the number of full adders in the reduction network. For the two-term 8-bit merged inner product element example, the two bit-product matrices are reduced through six stages of adders to produce a pair of intermediate operands which are summed with a single carry-lookahead adder. The total delay for this circuit T_{ma} is given by

$$T_{ma} = T_{gate} + 6T_{add} + T_{cla}(2M + \log k).$$

Digital Filter Case Study

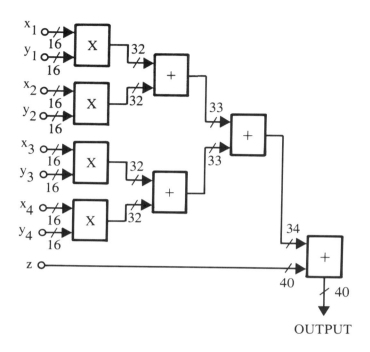

Figure 5-6. Expandable Four Term Multiplier Adder

Since carry-lookahead adders require interconnection topologies which are especially difficult to realize in VLSI circuits (because of the large number of signal crossovers), the saving of $2k - 2$ (k at the multiplier outputs and $k - 2$ in the adder tree) carry-lookahead adders in a k-term convolver is quite significant.

Another example will illustrate an even greater payoff for merged arithmetic: an eight-term FIR filter with an 18-bit expansion input is shown in Figure 5-7. The conventional implementation consists of eight multipliers and eight carry look-ahead adders as shown.

The merged implementation is shown in Figure 5-8. In this drawing, each entry is the height of the given column of bit products. Thus, the composite bit product matrix has nine bits in the least significant column, 17 bits in the next column, etc. The number of full and half adders used are listed in rows below each matrix. As shown on Table 5-2 the merged implementation requires only 73 percent as many two input gates as the conventional implementation requires.

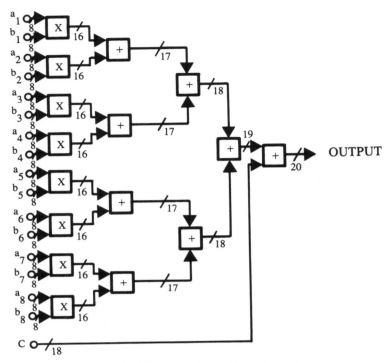

Figure 5-7. Multiterm FIR Filter Example

An apparent problem with this circuit is the large number of package pins required for the signals. A total of 166 signals were shown for the eight-term filter in Figure 5-7 (eight pairs of 8-bit inputs, an 18-bit expansion input, and a 20-bit result). In FIR filtering and most other applications, the vectors are elements of either a time sequence or are fixed kernels. In the first case, a single input port is required for each vector with shift registers to propagate the data. In the second case, the same shift register structure is used, but the kernel is loaded only at startup or when a new kernel is required. This reduces the data pin count to 62 (eight data vector input, eight data vector output, eight kernel input, 18 expansion input, and 20 for the result). Even with a normal assortment of control and power supply pins, this can be packaged easily in a standard 84 pin package.

The equivalent circuit of an expandable four-term FIR filter building block was shown previously in Figure 5-6. The output of the building block is the sum of the product x_1y_1, x_2y_2, x_3y_3, and x_4y_4 and the expansion input Z. Use of a latch between the output and the expansion input allows arbitrary expansion to the filter by

Digital Filter Case Study

1	1	1	9	17	25	33	41	49	57	65	57	49	41	33	25	17	9		COMPOSITE BIT PRODUCT MATRIX
																		65	FULL ADDERS
1	1	1	9	17	25	33	41	42	41	41	41	41	41	33	25	17	9		MATRIX II
				5F	13F	13F	13F	14F	13F	12F	11F	8F	3F					105	FULL ADDRESS
							1H	1H										2	HALF ADDERS
1	1	1	9	22	28	28	28	27	27	28	27	28	27	25	17	9			MATRIX III
				6F	9F	9F	9F	9F	9F	8F	8F	6F	3F					94	FULL ADDERS
1	1	1	15	19	18	19	19	18	18	19	18	18	19	17	9				MATRIX IV
			4F	6F	6F	6F	6F	6F	6F	5F	5F	4F	2F					74	FULL ADDERS
1	1	5	13	13	12	13	13	12	12	12	12	13	13	9					MATRIX V
		4F	4F	4F	4F	4F	4F	4F	4F	4F	4F	3F	2F					52	FULL ADDERS
1	1	9	9	9	8	9	9	8	8	8	8	9	9	9					MATRIX VI
		3F	3F	3F	3F	2F	3F	2F	2F	2F	2F	3F	3F	2F				41	FULL ADDERS
			1H					1H	1H	1H	1H							6	HALF ADDERS
1	4	6	6	6	6	6	6	6	6	6	6	6	6	5					MATRIX VII
1F	2F	2F	2F	2F	2F	2F	2F	2F	2F	2F	2F	2F	1F	1F				31	FULL ADDERS
2	4	4	4	4	4	4	4	4	4	4	4	4	4	3					MATRIX VIII
1F	1F	1F	1F	1F	1F	1F	1F	1F	1F	1F	1F	1F	1F	1F				17	FULL ADDERS
3	3	3	3	3	3	3	3	3	3	3	3	3	2	1					MATRIX IX
1F	1F	1F	1F	1F	1F	1F	1F	1F	1F	1F	1F	1F	1F					16	FULL ADDERS
													1H					1	HALF ADDER
1	2	2	2	2	2	2	2	2	2	2	2	2	2	1					MATRIX X
						17 STAGE CARRY LOCKAHEAD ADDER													
						RESULT													

Figure 5-8. Merged 8-Term FIR Filter Matrix Reduction

repeatedly accumulating four terms at a time. Beyond its application to FIR filtering, this merged multiplier-adder can be used effectively to implement the complex multiplication which is widely used in signal processing.

		Conventional		Merged	
Function	Gates	Usage	Gates	Usage	Gates
AND Gate	1	512	512	512	512
Half Adder	4	56	224	9	36
Full Adder	20	280	5600	495	9900
14 Bit CLA	515	8	4120		
16 Bit CLA	515	4	2060		
17 Bit CLA	673	2	1346	1	673
18 Bit CLA	673	1	673		
19 Bit CLA	673	1	673		
Total Gate Count			15208	(73%)	11121

Table 5-2. Eight-Term 8-Bit FIR Filter Implementation Comparison

The four-term filter building block is used much like the commercial multiplier-accumulator. The first four terms of a given filter output are computed and latched into the accumulator. The next four terms are computed and summed to the accumulator contents, etc.

Given the advanced 1 micron design rule bipolar technology, the merged arithmetic building block can operate at a 20 MHz clock rate. Since four terms are computed simultaneously, a system performance of 8×10^7 filter terms per second is achieved.

Sign-Logarithm Implementation

Over the years, a wide variety of number systems have been proposed for application in special-purpose computers, of which a significant fraction have actually been exploited in computer arithmetic units [5-7]. Most of the practical implementations have used variations of the familiar weighted binary number systems (e.g., binary fractions, binary integers, or binary floating-point numbers) in spite of two major problems inherent in all of these number systems: 1) the relatively low efficiency (i.e., speed/complexity) for multiplication, and 2) the requirement for large data word size to satisfy dynamic range requirements.

Digital Filter Case Study

This section develops a design for an FIR filter based on the sign/logarithm number system [5-8]. This approach circumvents the low speed of multiplication with conventional weighted binary arithmetic while avoiding the difficulty of scaling in residue number systems. It is not appropriate for general-purpose computation, but is suitable for special-purpose computation in many signal processing applications.

The sign/logarithm representation of an integer operand consists of the sign of the operand appended to the logarithm of the absolute value of the operand. It is possible to use this approach to represent fractions or mixed numbers by inclusion of a scale factor which may be varied on an individual basis or on a block basis. The number system and algorithms for the three basic signal processing operations, addition, subtraction, and multiplication, are briefly described to illustrate the implementation simplicity which can be achieved in the design of specialized processors.

For the sign/logarithm number system, any integer A is represented by its sign S_A and the binary logarithm L_A of its magnitude:

and

$$\left. \begin{array}{c} S_A = 1 \quad \text{if } A \leq 0 \\ S_A = 0 \quad \text{if } A \geq 0 \\ (\text{note that } S_A = 0 \text{ or } 1 \text{ if } A = 0) \\ L_A = \log(|A|) \quad \text{if } |A| > 1 \\ L_A = 0 \quad \text{if } |A| \leq 1. \end{array} \right\} \quad (5\text{-}4)$$

All computations in the number system are performed on the "signed logarithm" Σ_A which represents A as

$$\Sigma_A = (1 - 2S_A)L_A.$$

The relationship between Σ_A and A is sketched in Figure 5-9 for $-64 \leq A \leq 64$ for 4-bit sign/logarithms. It is apparent from the graph that comparison of the magnitudes of various numbers in implemented as easily for signed logarithms as for the original number system since the logarithm is a monotonic function. The number A may be found from Σ_A by separating the latter into S_A and L_A and applying the relationship.

$$A = (1 - 2S_A)2^{L_A}.$$

In the examples which follow, procedures are described to generate a signed logarithm result by performing each of the basic arithmetic operations on a pair of sign/logarithm operands (S_A, L_A) and (S_B, L_B). Since the algorithms for addition

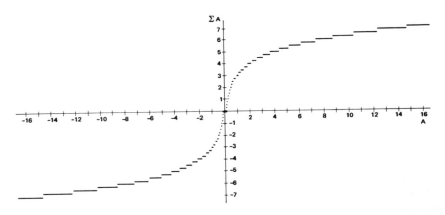

Figure 5-9. Sign/Logarithm Number System

and subtraction are based on a multiplicative procedure, the multiplication algorithm is described first.

The logarithm of the product of two operands in the sign/logarithm system is simply the sum of their logarithms:

$$L_P = L_A + L_B = \log(A) + \log(B) = \log(AB) \qquad (5\text{-}5)$$

As with conventional sign/magnitude number systems, the sign of the product is determined by modulo-2 addition of the signs of the factors:

$$S_P = S_A \oplus S_B$$

in which the symbol $\oplus$ indicates the Exclusive-Or operation.

Since multiplication involves the addition of two logarithms, it is possible for overflow to occur. In the overflow conditions, the magnitude of the result exceeds the largest numer which can be represented with an n-bit binary logarithm in a manner directly analogous to overflow in a weighted binary number system; corrective action can be initiated on its detection if desired. Thus, unlike the residue system, the sign/logarithm system allows easy detection (and correction, if desired) of overflow conditions.

The algorithms for addition and subtraction in the sign/logarithm number system are based on an extension of multiplication [5-9]. Consider the sum S of two operands A and B:

$$S = A + B.$$

Digital Filter Case Study

Rearranging yields

$$S = A(1 + B/A). \qquad (5\text{-}6)$$

Here, the sum is written as a product of the first number A times a factor (1 + B/A) which is a function of the ratio of the two operands. This result can be restated in the form.

$$S = AF(B/A)$$

where $F(X) = 1 + X$. Three operations are therefore required to compute the sum of A and B. First, the ration B/A is computed, then the function F is evaluated, and finally the multiplication of A by the value of the function F is performed.

Extending the above formulation to operations in the domain of sign/logarithm numbers produces the following procedure. Note that addition or subtraction of operands of arbitrary sign is identical, except that different tables of the function F are used depending on whether the signs are alike or different:

$$L_S = L_A + \beta(L_B - L_A)$$
$$L_D = L_A + \gamma(L_B - L_A) \qquad (5\text{-}7)$$

where $\beta(X) = \log(1 + 2^X)$ and $\gamma(X) = \log(1 - 2^X)$, L_S is the logarithm of the sum, and L_D is the logarithm of the difference.

The physical realization of a sign/logarithm adder/subtractor with a binary adder, a binary subtractor, a Read-Only Memory (ROM), and a small amount of ancillary logic is diagrammed in Figure 5-10. The ROM for β and γ each contain 2^{N+1} words of N bits for a total of $N2^{N+2}$-bits of storage. For N = 8, a total of 8192 bits are required, which is easily implemented with a single 1024 × 8 ROM.

As an extension of the fundamental operations of addition, subtraction, and multiplication in sign/logarithm arithmetic, a multiterm sign/logarithm convolver can be configured to implement the multiplication and summation required by the FIR filter. The detailed design is depicted in Figure 5-11 for a two-term case with two sign/logarithm multipliers followed by two sign/logarithm adders.

The total complexity of a two-term FIR filter element, assuming N-bit multipliers and multiplicands and an N + 2-bit expansion input, is determined by inspection to be

2	N-bit adders
2	N + 1-bit adders
2	N + 2-bit adders
1	$2^{N+3} \times$ (N + 1)-bit ROM
1	$2^{N+4} \times$ (N + 2)-bit ROM.

The total propagation delay of the two-term convolution element is that required for five full-word adders (a total of 5N + 6 full adder delays if simple ripple carry adders are used) and the access delay of two ROMs.

Although the sign/logarithm implementation of direct convolution is an efficient design, overall device complexity is a strong function of the required word size N. Specifically, as the word size increased by one bit, the length of each of the six adders increase proportionally, and the required amount of ROM more than doubles. For 6- to 8-bit word size, the total amount of ROM storage required is quite reasonable, but increased rapidly with increasing word size, and is currently unattractive for N > 12.

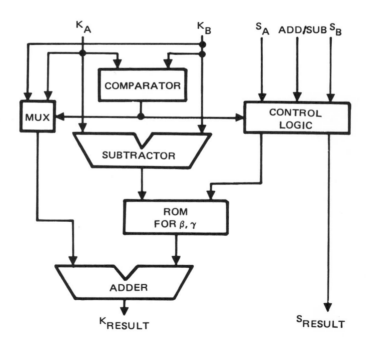

Figure 5-10. Sign/Logarithm Adder/Subtractor

Digital Filter Case Study

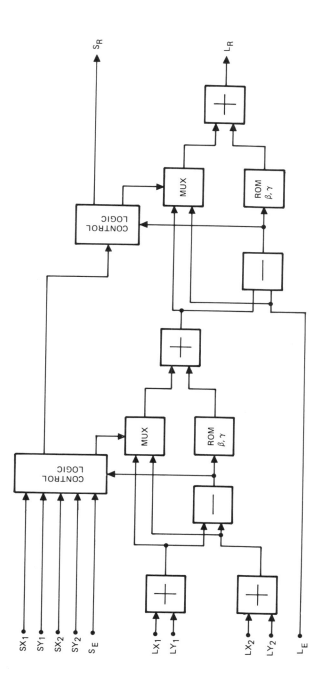

Figure 5-11. Two-Term Sign/Logarithm FIR Filter

The primary motivation for specifying large word sizes in signal processing and special-purpose computational applications is the requirement to accommodate a wide dynamic range of signals. If the relative dynamic range of the data operands is defined as the range (i.e., the difference between the most positive and the most negative numbers), divided by the step size for small signals (i.e., the step between the two positive levels closest to the origin), then the relative dynamic range for an N-bit two's complement binary number may be specified as $D_b(N)$:

$$D_b(N) = 2^N - 1,$$

whereas the relative dynamic range of a K-bit sign/logarithm number $D_1(K)$ is

$$D_1(K) = 2^{2^{(K-1)}} - 1.$$

Comparison of the equations for D_b and D_1 indicate that the dynamic range of a K-bit sign/logarithm number exceeds the dynamic range of an N-bit two's complement binary number in this sense when

$$2^{K-1} - 1 \geqslant N.$$

From this relation, when $N = 8$, $K = 5$ and when $N = 16$, $K = 6$. Thus, a 6-bit sign/logarithm number will produce a dynamic range, in the sense defined above, better than that of a 16-bit binary number. A sign/logarithm implementation of a direct convolution unit employing 6-bit operands (that is, with $K = 6$) requires only 42 adders, assuming that ripple carry addition is used to perform the arithmetic, and 512×7-bit and $1{,}024 \times 8$-bit ROM's. This is a total complexity of 840 gates of "random logic" and 11,776 bits of ROM. At an equivalence factor of two ROM bits per gate, the total complexity is 6,728 gates.

An eight-term sign/logarithm filter building block can be developed with the one micron design rule bipolar technology. The resulting circuit will operate at a 20 MHz clock rate giving a system performance of 1.6×10^8 filter terms per second. This building block is used like the merged arithmetic four-term filter to implement FIR filters, in that it computes the equations of (5-2) on a row at a time basis.

Modular Array Implementation

The modular array represents a case where the arithmetic operators exhibit a high degree of latency. Latency is described in Figure 5-12. It occurs when multiple sets of data enter a computational element prior to the emergence of the first result (i.e., the time to execute a single computation is greater than the data entry cycle time). Latency may be modeled as a single cycle computational element connected to a multicycle delay. For example, a long latency multiplier takes in data at a very high rate but requires a number of clock cycles to complete the multiplication. It can continue to take in operands during the time that it is computing the first product but it will take multiple cycles to generate each result.

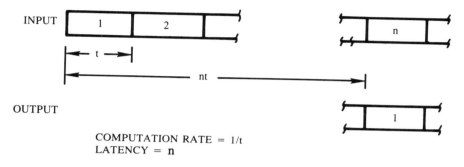

Figure 5-12. Latency Definition

High latency multipliers can be used to implement the FIR filter by starting the calculation of X(0) times K_1, while that is in process starting X(-1) times K_2, while that is in process starting X(-2) times K_3, etc. All the data entering the multiplier are processed in order, and the results are available in order. To sum the products, the first two can be entered into a high latency adder, but to accumulate the sum of multiple products is a problem since the first sum doesn't exit the adder until many clock cycles later. The produces gaps during which the adder is inactive.

An alternative approach is to process the data in an "input driven" manner where all filter terms involving a given sample are computed and stored in partial accumulators. For example assume X(0) has just been received. The products X(0) times K_1, X(0) times K_2, X(0) times K_3, X(0) times K_4 . . ., X(0) times K_{n-1} are computed. These terms are added into registers that are forming $Y_1, Y_2, Y_3, Y_4, \ldots, Y_{n-1}$. This results in a configuration with n filter values being processed simultaneously by the use of n registers, each holding a partial result. Effectively the equations of (5-1) are being evaluated along the diagonal in contrast to the row wise approach used by the other three filter implementations of this section.

The basic approach here is to develop a circuit, called the modular array, that can be connected into an array to build very high speed adders, subtractors, and multipliers. The modular array circuit is designed to operate a clock rates of up to 500 MHz by pipelining, although the latency (the time from when the operands are entered until the result is available) will be long relative to the clock rate (on the order of tens of nanoseconds). In the multiterm convolver, as well as many other signal processing applications, large block of data are processed without data-dependent branch instructions so that the processing rate is more important than the latency. The modular array circuit is a carry-save implementation of Booth's multiplication algorithm [5-10] which has been widely accepted for both software and hardware multiplication applications. As noted recently in [5-11], this algorithm may be pipelined using a carry save approach for the addition/subtraction of the multiplicand to the partial product. For example, a 16×16 multiplier requires 32-stage pipeline; the first 16 stages of the pipeline form two intermediate results (i.e., a sum word and a carry word) which are added by the final 16 stages to form the product. Since the pipeline clock interval need only be long enough to permit execution of a carry-save addition operation between successive clock cycles, pipeline rates on the order of 500 MHz can be readily attained with advanced oxide isolated ECL process technology. Booth's algorithm was selected as the basis for the design of the modular array circuit since a single VLSI circuit design can be used for multiplication (by using the complete multiplier configuration), or for addition or subtraction by using only the carry-save adder array.

In Booth's multiplication algorithm, the bits of one operand (the multiplier) are examined in overlapping pairs, and a decision is made whether to add or subtract the other operand (the multiplicand) from the accumulated sum of partial products or to retain the previous sum. The pair of multiplier bits are inspected; if both multiplier bits are the same (either 00 or 11), the partial product is retained. If the multiplier bit pattern is 10, the multiplicand is subtracted from the partial product; if the bit pattern is 01, the multiplicand is added to the partial product. Finally, the partial product and the multiplier are shifted right by one bit position and the next pair of multiplexer bits are inspected. This procedure is repeated N times for an N-bit multiplier. The requirement for physical realization of this algorithm is an array of full adders and a small amount of peripheral logic, as depicted in Figure 5-13. Algebraic subtraction is performed when required by adding the two's complement of the multiplicand; this is implemented directly within the device by first forming the one's complement, and then by adding a 1 bit at the least significant bit position.

The carry-save adder inputs the sum and carry bits from a previous row (or level) of the adder matrix into the next lower level of adders, thereby eliminating horizontal carry propagation within a row of the full adder matrix. This results in an array with maximum pipeline rate since only a single adder operation must be completed within each clock period.

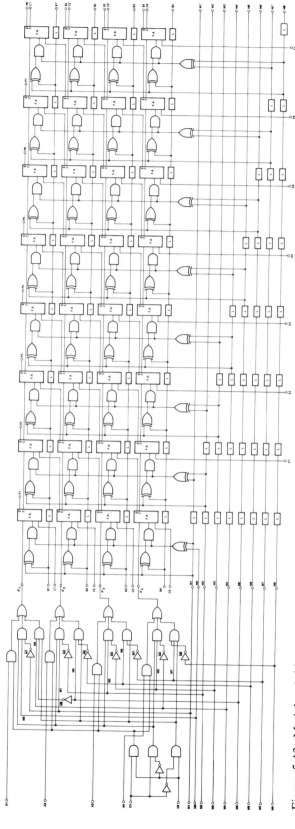

Figure 5-13. Modular Arithmetic Array

As shown in Figure 5-13, a single integrated circuit implementating this modified form of Booth's algorithm consists of three sections: a triangular array of pipelining latches to hold the multiplier word, a rectangular array of pipelining latches to hold the multiplicand, and an array of full adders supported by a small amount of ancillary logic. The triangular array of latches on the left side appropriately delays the multiplier bits, corresponding to the delay encountered by the multiplicand bits and partial products during their propagation through the carry-save adder circuit. Thus the pair of multiplier bits that determine whether to add or subtract the multiplicand to the partial product are properly aligned with the multiplicand and the partial product. The inclusion of pipelining latches at every level of processing throughout the array ensures maximum operating speed.

The modular arithmetic array may be used to implement a variety of operations, including addition, subtraction, and multiplication. The simplest designs generate the results on a "time skewed" basis, that is, the resulting bits are not all present at the output simultaneously, but emerge from the unit sequentially in order from the least significant to more significant bit positions as they are formed. For example, the least significant result bit is available one clock period after the operands are presented to the device inputs. The next most significant bit is available one clock period later, and so on. If necessary, the skewed result can be deskewed by using the triangular array of latches.

The resulting chip is a four wide by eight deep array, with 32 full adders, and with a total complexity on the order of 80 gates requiring 56 signal pins (which is consistent with the limitations on the oxide isolated ECL technology as of 1980 when the design was completed.)

Table 5-3 indicates the number of modular array circuits that are required to implement various size adders and multipliers with deskewed results. These adders and multipliers are the building blocks that will be used to implement the FIR filter as shown on Figure 5-14. The kernel values are held in a shift register memory. For each data value entered into the upper multiplier input port, the kernel shift register is circulated fetching all kernel values into the other multiplier port. The 16×16 multiplier with its 32 clock latency feeds the 40 bit adder with its 40 clock latency. The adder outputs are accumulated in the output shift register thereby implementing the "input driven" scheme.

Although at first consideration, a pipelined adder with a 40 clock period latency appears unsuitable for use in evaluating FIR filters, highly efficient operation can in fact be obtained by interleaving $N (\geq 40)$ filter outputs and computing a new component for each of the filters every N clock cycles. This process is

Digital Filter Case Study

shown in Figure 5-14. For each input value, $X(i)$, n kernel values $(K_1, K_2, \ldots, K_{n-1})$ are accessed and entered into the multiplier to form n products which become available after the multiplier latency time. These products are then summed in the accumulator, adding to each of the n FIR calculations which are in progress. During each cycle, one filter value will be completed and another will be initiated in its place.

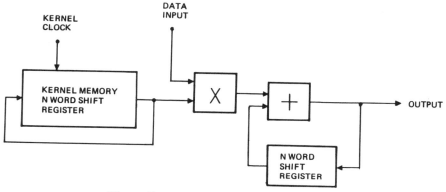

Figure 5-14. Modular Array FIR Filter

Function	Word Size	Number of 4×8 Modular Arrays
Adder	8	2
Adder	16	6
Adder	24	12
Adder	32	20
Adder	40	30
Multiplier	8	6
Multiplier	12	12
Multiplier	16	20

Table 5-3. Arithmetic Implementation with the Modular Array

To implement the FIR filter, a single high-speed multiplier/accumulator is used. With 16-bit operands producing a 32-bit product and a 40-bit filter output, 256 point filters can be computed without overflow. A total of 50 modular arrays are required, i.e., 20 for the multiplier and 30 for the adder. This assembly achieves a computational capacity of 5×10^8 filter terms per second (i.e., a throughput of 10 million filter terms per second, per circuit).

In summary, the modular array circuit represents an unconventional approach to the development of a logical array which is realized via three separate (interacting) iterative functions. This approach contrasts to the more conventional approach in which many identical cells are interconnected to form a two-dimensional array. The advantage of the modular array circuit is that a single design can be used efficiently as either an adder/subtractor or as a multiplier. This circuit is well suited to signal processing applications in which its relatively long pipeline length and attendant latency is not a limitation, in contrast to general-purpose data processing applications where the long output delays could pose substantial problems due to the frequent need for data-dependent branch instructions.

5.4 FIR Filter Comparison

Formal comparisons of the four implementations proposed here for FIR filter processors must consider several crucial factors, particularly the state of the art of VLSI circuit technology, which impacts both circuit speed and circuit density. Second, the choice of the number system for the arithmetic processor is critical, since it affects the dynamic range, accuracy, and word size of the hardware implementation. Although the approaches considered here are implemented in three different technologies characterization of their speed/complexity performance can be performed on the basis of the throughput (here the number of filter terms computed per second) divided by the implementation complexity. A variety of complexity measures have been used for specific applications, but one of the most useful is the number of integrated circuits required to implement the design. This measure corresponds well with more conventional system complexity measures (e.g., cost, size, power, etc.), is easily computed, and is useful in preliminary feasibility assessments. For purpose of this comparison all designs provide 16-bit dynamic range. A large numerical value of the figure of merit implies increased computation per integrated circuit, or alternatively, fewer circuits to achieve equivalent computation rates.

As shown in Table 5-4, the sign/logarithm convolution unit exhibits a figure of merit of 160, which is twice that of the merged arithmetic element and an order of magnitude greater than that of the modular array circuit and the commercial multiplier accumulator. Comparison of the approaches which employ conventional two's complement arithmetic demonstrates that the merged arithmetic element implemented with high-density moderate-speed VLSI technology is more efficient than the ultra-high-speed modular array circuit. This finding corroborates one of the major motivations for development of VLSI technology: is consistent with the intuitive notion that increases in circuit density reduce partitioning "breakage" since a complete function is performed by a single circuit.

Approach	Chips	Terms per Second	Merit*
Modular Array	50	500M	10
Merged Arithmetic	1	80M	80
Commercial Multiplier-Accumulator	1	8M	8
Sign/Logarithm	1	160M	160

Table 5-4. FIR Filter Implementation Comparison

The higher figure of merit for the implementation using sign/logarithm arithmetic relative to two's complemnt merged arithmetic demonstrates the potential advantage of unconventional arithmetic. Recognizing that the sign/logarithm error characteristics are not yet well understood, detailed simulation studies are required to ensure that the constant relative error associated with this specialized approach does not introduce drawbacks which outweigh the gain in performance/complexity.

In summary, this case study demonstrates the basic approach involved in developing a signal processing system. The FIR filter is widely used and serves as a convenient example since a variety of approaches can be exploited in its implementation. Of the four approaches considered, the advanced one micron design rule bipolar technology provided substantially better system performance than either the commercial biploar (two micron design rule) or the very fast ECL technology. The latter approach suffers in this comparison largely due to "functional breakage" since the multipliers and adders have to be partitioned into many circuits as a result of complexity limitations.

5.5 References

5-1 Earl E. Swartzlander, Jr. and Barry K. Gilbert, "Arithmetic for Ultra High-Speed Tomography," **IEEE Transactions on Computers,** Vol. C-29, 1980, pp. 341-353.

5-2 L. Dadda, "Some Schemes for Parallel Multipliers," **Alta Frequenza,** Vol. 34, 1965, pp. 349-356.

5-3 Earl E. Swartzlander, Jr., "Merged Arithmetic," **IEEE Transactions on Computers,** Vol. C-29, 1980, pp. 946-950.

5-4   Charles R. Baugh and Bruce A. Wooley, "A Two's Complement Parallel Array Multiplication Algorithm," **IEEE Transactions on Computers,** Vol. C-22, 1973, pp. 1045-1047.

5-5   P.E. Blankenship, Comments on "A Two's Complement Parallel Array Multiplier Algorithm," **IEEE Transactions on Computers,** Vol. C-23, 1974, p. 1327.

5-6   William J. Stenzel, William J. Kubitz, and Gilles H. Garcia, "A Compact High-Speed Parallel Multiplication Scheme," **IEEE Transactions on Computers,** Vol. C-26, 1977, pp. 948-957.

5-7   Harvey L. Garner, "Number Systems and Arithmetic," in Franz L. Alt and Morris Rubinoff, eds., **Advances in Computers,** New York: Academic Press, 1965, pp. 131-194.

5-8   Earl E. Swartzlander, Jr. and Aristides G. Alexopoulos, "The Sign/Logarithm Number System," **IEEE Transactions on Computers,** Vol. C-24, 1975, pp. 1238-1242.

5-9   N.G. Kingsbury and P.J.W. Rayner, "Digital Filtering Using Logarithmic Arithmetic," **Electronics Lettes,** Vol. 7, 1971, pp. 56-58.

5-10 Andrew D. Booth, "A Signed Binary Multiplication Technique," **Quarterly Journal of Mechanics and Applied Mathematics,** Vol. 4, 1951, pp. 236-240.

5-11 J. Robert Jump and Sudhir R. Ahuja, "Effective Pipelining of Digital Systems," **IEEE Transactions on Computers,** Vol. C-27, 1978, pp. 855-865.

Frequency Domain Filter Case Study

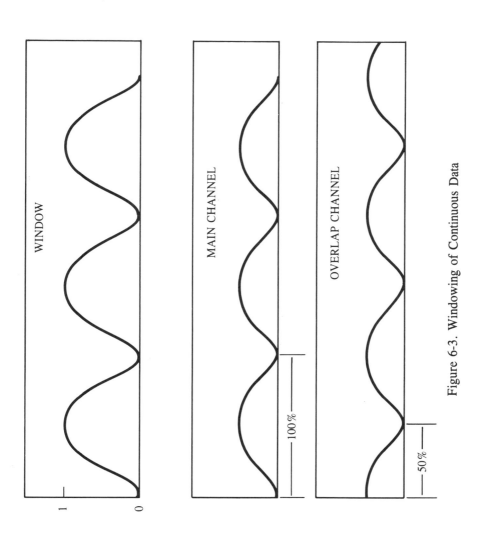

Figure 6-3. Windowing of Continuous Data

4,000. Since an FFT algorithm will be used, the length should be a power of two so a transform length of 4,096 points is used. The data is continuous, which means that it is necessary to use two data streams operating with overlap. Floating point arithmetic was selected for this system to simplify data scaling and to eliminate data overflow/underflow problems.

6.2 Technology Selection

An important aspect in exploiting technology to the fullest, is clear identification of the optimization criterion. This varies depending on application (i.e., minimizing the parts count, power, size, cost, number of part types, etc.). For this example, the goal is to minimize the parts count, which is achieved by maximizing the computation per part. The computation per part is estimated by the product of the number of gates times their clock rate. A rough estimate of the computation per part as a function of clock rate for commercially available parts in 1985 is shown in Figure 6-4.

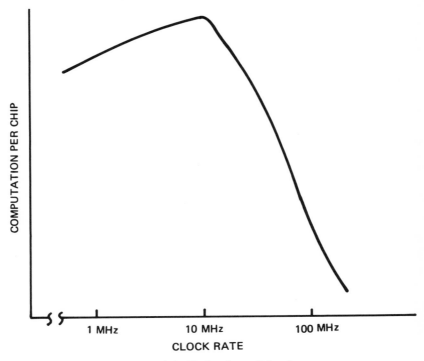

Figure 6-4. Technology Selection

It is evident that the best computational density is achieved with clock rates on the order of 10 MHz. At higher rates, ECL interfaces become mandatory to achieve the speed, but the density of the devices is much lower than the gain in speed, and there is a very limited selection of functions. Given a constraint of commercially available devices, it is generally better to use TTL running at 10 MHz than to use ECL running at 100 MHz or even faster. At speeds below 10 MHz most commercially available circuits are early generation low density MOS devices. Thus from either side there is a benefit in moving towards the 10 MHz rate for the clock. This phenomenon is highly time dependent. If the curve were developed ten years ago, the peak would have been around 1 MHz. It would have been much lower as well. The optimum clock rate has been increasing by about an order of magnitude per decade. This trend will likely continue in the forseeable future. Clock rates on the order of 20 to 40 MHz are being achieved for high density chips on the VHSIC program and eventually VHSIC like commercial chips will be available.

Another way of viewing the technology choice is to examine the level of functional complexity attainable on a single VLSI circuit. For most functions, there is a "logical critical mass" required for implementation. At lower logic levels the function can not be implemented on a single chip and must be partitioned for either multiple chip implemenation (which involves extra I/O drivers, package pins, etc.) or for sequential implementation (which requires additional control logic, temporary variable storage, etc.). Clearly, the implementation efficiency is greatly improved if the complete function can be realized on a single chip. As a result of these considerations, a 10 MHz clock rate was selected.

6.3 Frequency Domain Filter Implementation

To achieve the desired rate of 40 MHz with circuits operating at a 10 MHz clock rate, an internal parallelism of four is used. Of the modules shown on Figure 6-2, the FFT and inverse FFT are the critical elements as they are roughly an order of magnitude more complex than the other modules. Accordingly, their implementation receives the greatest emphasis, although the implementation of each of the modules will be briefly described for completeness.

Data Acquisition Module

This module interfaces to the data source, produces the four data channels from the single input data stream, and windows the data in the time domain. Its implementation is shown in Figure 6-5. Complex data is clocked at a 40 MHz rate

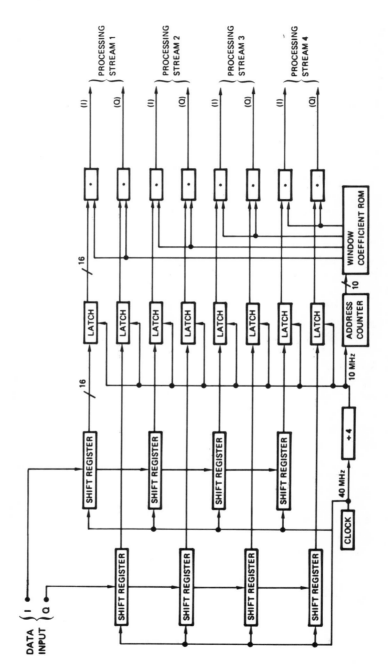

Figure 6-5. Data Acquisition Module

into 4 pairs of shift registers (pairs of registers are used to accommodate the two components of each complex data word). The contents of the shift registers are then transferred into 8 parallel latches at a 10 MHz rate. This shift register/latch combination converts the single fast serial channel to four slower parallel channels. The data are windowed by multiplying with window coefficients that are stored in a Read Only Memory (ROM).

On this module the shift registers and latches are implemented with ECL devices in order to achieve the 40 MHz input clock rate. The multipliers and ROMs are commercial VLSI circuits. Since commercial ECL devices are quite low in density over 80% of the 120 chips required to implement this module are ECL circuits.

The input data are assumed to be 16-bit fixed-point words. Since most high speed A/D converters are lower in precision (i.e., 8 bits, 10 bits, or occasionally as high as 12 bits) it may be possible to use smaller data words in this module. Because of the low level of integration of the ECL components, reduction to 12 or 8 bit input data word sizes reduces the complexity by nearly 25% or 50% respectively.

FFT Architecture

In surveying FFT algorithms for a high speed system, the pipeline FFT algorithm [6-2] - [6-5] is especially attractive. As shown in Figure 6-6 it consists of a number of butterfly computational units interspersed with delay commutator elements that provide interstage data reordering. With this system it is necessary to select the appropriate radix for the butterfly computation. A radix R butterfly takes in R data values, multiplies all but one of them by complex sine and cosine weights, adds and subtracts them from each other, and generates R intermediate results.

RADIX R BUTTERFLY

DATA RATE IS R TIMES THE CLOCK RATE

DELAYS BETWEEN BUTTERFLY COMPUTATION ELEMENTS PROVIDE INTERSTAGE REORDERING

Figure 6-6. Pipeline FFT Processor Architecture

The data rate is R times the clock rate. Thus with a radix 2 system operating at a clock of 20 MHz, data is processed at a 40 MHz rate because two data are processed on each clock period. Similarly with radix 4 and 10 MHz clock rates, the processing rate is 40 MHz, etc. With a 10 MHz clock speed as dictated by technology considerations, use of a radix 4 pipeline FFT is adequate to achieve the desired data rate of 40 MHz.

The pipeline FFT structure is shown in more detail in Figure 6-7. It is contructed with two type of building blocks, radix 4 butterfly computational elements and programmable length delay commutators. They are interspersed in pipeline form, first a computational element then a delay commutator, then a computational element, etc.

FFT Computational Element

The radix 4 butterfly computational element realizes a 4 point discrete Fourier transform. It is implemented with 3 complex multipliers, 8 complex adders (4 performing addition and 4 performing subtraction), cosine/sine tables (stored in ROM), and miscellaneous addressing logic as shown on Figure 6-8. The computational element is implemented with approximately 80 integrated circuits. This low level of complexity is a direct result of the commercial availability of floating point arithmetic components realized as single VLSI circuits. Due to the importance of the arithmetic components, a brief description of the characteristics of typical single VLSI floating point circuits is included here. See [6-6] for further information on 22-bit floating point adders and multipliers.

Adder: Under user control, the adder performs floating point addition, accumulation, and conversion between fixed and floating point formats. Rounding and scaling (i.e., $\div 2$) are also selectable if desired. For proper operation and maximum accuracy, non-zero floating-point operands must be normalized, with fractions in the range of $-1.0 \le F < -0.5$ or $0.5 \le F < 1.0$.

The adder performs the three component operations of floating point addition: denormalization (exponent alignment), addition, and renormalization. The first and last steps are hardware intensive, involving shifting the fraction and incrementing the exponent by the number of shifts performed. In the addition mode, the adder first selects the addend with the smaller exponent. This operand is denormalized by shifting it rightward by the difference between the exponents of the two addends. The denormalized fraction is added to fraction of the other addend and their sum passes to the renormalizing section, along with the exponent of

Frequency Domain Filter Case Study

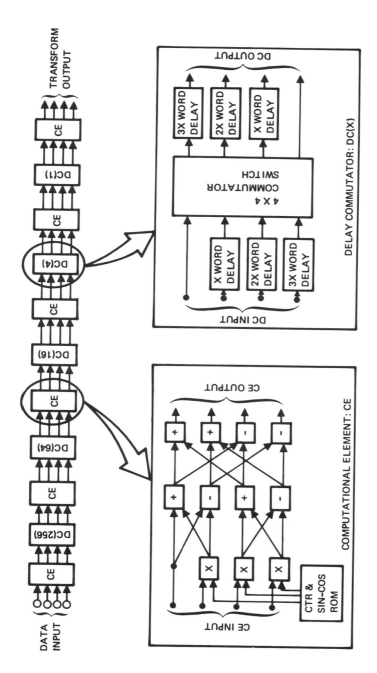

Figure 6-7. Radix Four Pipeline Cooley-Tukey FFT

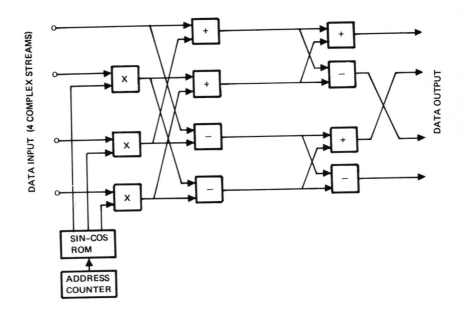

Figure 6-8. Radix Four Computational Element

the larger addend. The sum is normalized by shifting its fraction leftward until the sign bit differs from the next bit, and the exponent is decremented by the number of bit positions of this shift.

Subtraction is identical to addition, except that the fraction of the subtrahend is complemented before the addition is performed. In standard two's complement fashion, the bits are inverted and a "hot one" is introduced at the adder's LSB carry-in position. Fixed-to-floating-point conversion and normalization of floating point numbers can be performed by left shifting the fraction as necessary to eliminate redundant leading zeros or ones while decrementing the exponent to compensate.

Multiplier: The floating point multiplier is basically a two's complement fixed point multiplier for the fractional part of the floating point word, a small adder for the exponent of the floating point word, and a simple normalizer. Its output conditioning requirements are minimal: if the input operands are normalized, then the product is at most one shift left or right from normalization. Hence, the barrel shifter used to normalize the adder result is replaced by a small 3-position multiplexer. The only data dependent interaction between fraction and exponent occurs in the final product normalization step, where the exponent must be incremented or decremented to compensate for any shift in the fraction. With the normalizer defeated, the chip performs two's complement, fixed point multiplications.

FFT Delay Commutator

The interstage data reordering required at stage i in the implementation of a 4^n point transform is a base 4 digit reversal of the data elements in a 4×4^n matrix. This is performed with a "delay commutator." As shown on Figure 6-9 the delay commutator consists of three input delay lines, a commutator switch, and three output delay lines. Data enter the delay commutator through four parallel channels. The first data path receives no delay, the second receives a delay of 4^{n-i-1}, the third receives a delay of $2 \times 4^{n-i-1}$, and the fourth receives a delay of $3 \times 4^{n-i-1}$. The data then passes through the commutator switch, where they are switched to selected data paths. Lastly, the data are deskewed through a second set of delay lines.

The routing of data that occurs in processing a 64 point transform with the radix 4 pipeline FFT algorithm is graphically charted in Figure 6-10, after [6-7, p. 611]. Input data numbered by their time order of arrival are shown on four parallel streams at the top of the figure. These data are processed by a computational element producing intermediate results with the same time sequencing. The butterfly output are processed through a delay commutator with $X=4$ as shown at the top of the figure. The delay commutator rearranges the four streams of data separated by 16 points into four streams where the data are separated by four points. These data are then operated upon by a second computational element which does not change the data order. A second delay commutator set for $X=1$ reorders the data to produce streams of adjacent data.

Implementation of this delay commutator with commercially available circuits is quite difficult because delay lines of many different lengths are required. For example, for a 4K FFT delay lengths of 1, 2, 3, 4, 8, 12, 16, 32, 48, 64, 128, 192, 256, 512, and 768 are required. Commercially available devices include few delay lines and none with the appropriate speeds and lengths.

A complete 4K FFT requires 11 elements. There are 6 computational elements with 80 commercial integrated circuits each and 5 delay commutators with 180 circuits each for a total of 1380 circuits.

130 VLSI Signal Processing Systems

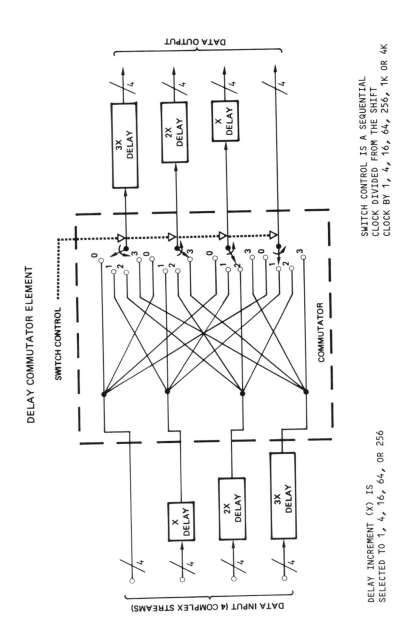

Figure 6-9. Delay Commutator Element

Frequency Domain Filter Case Study

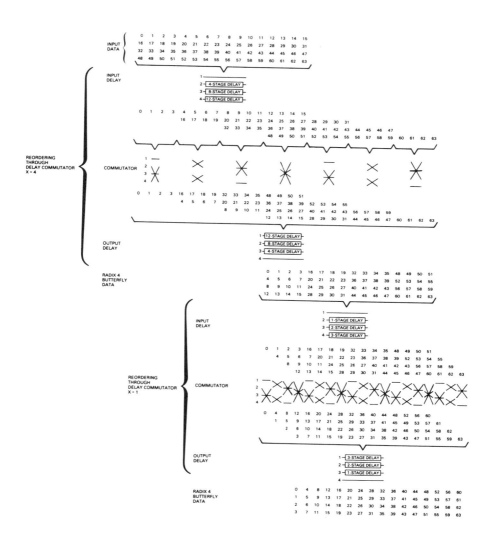

Figure 6-10. Data Patterns Through Two Delay Commutators for a 64 Point FFT. After [6-7, p. 611]

Power Spectral Density Module

The power spectral density module is used to compute an estimate of the power spectral density, PSD, by summing the squares of the real and complex components of each frequency of the spectrum. It also generates an average of successive PSD using a recursive averaging scheme. The design of the element is shown in Figure 6-11. The magnitude of the complex input data is computed followed by the averaging operation. The averaging involves summing the current spectrum with a fraction of the previously accumulated spectrum on a point-by-point basis. The complexity of the Power Spectral density module is approximatly 90 integrated circuits.

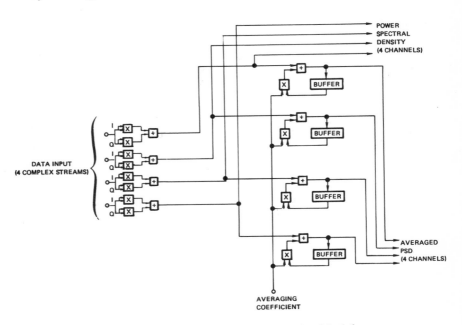

Figure 6-11. Power Spectral Density Module

Freqency Domain Filter Module

The frequency domain filter multiplies the 4096 point complex spectrum by the filter kernel on a point by point basis. The module is implemented with four identical complex channels, as shown if Figure 6-12. The filter kernel is stored in a RAM for recall as input data arrives. This module is implemented with approximately 70 integrated circuits. The filter kernel is computed by a general

Frequency Domain Filter Case Study

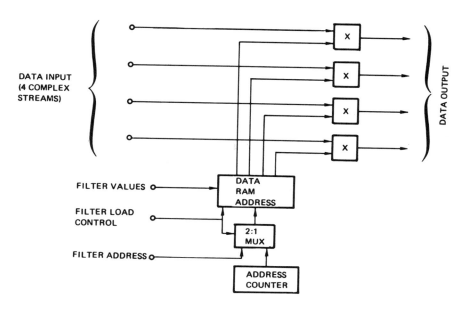

Figure 6-12. Frequency Domain Filter Module

purpose mini or microcomputer. The filter response can be fixed or made adaptive depending on the system application. The general purpose computer is not included in the filter complexity estimates since most systems have a host computer that will be used to provide this function.

Inverse FFT Module

The inverse FFT is structured exactly like the FFT. The differences are that the values of the sine and cosines are rescaled and the ordering of the delay commutators is reversed as shown in Figure 6-13. The implementation is a direct rearrangement of the basic modules.

Output Combiner Module

The output combiner shown on Figure 6-14 converts the four TTL data streams to ECL logic levels. It then multiplexes the data through four to one multiplexers that form the proper sequence of time domain signals and then through a two to one multiplexer that selects either the normal or the overlap channel.

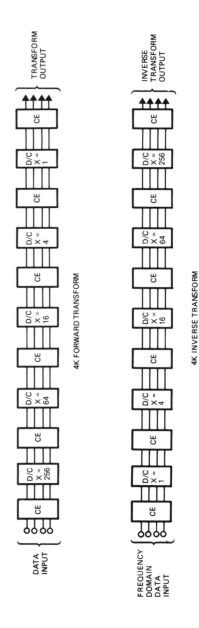

Figure 6-13. Forward and Inverse Pipeline FFT Implementation

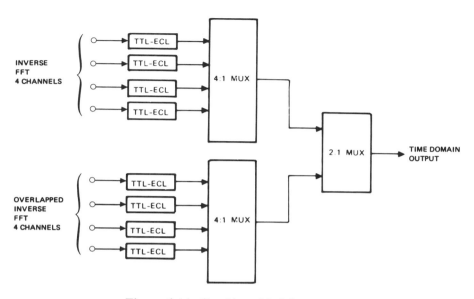

Figure 6-14. Combiner Module

Filter Complexity Summary

The complexity of the complete two channel frequency domain adaptive filter is summarized in Table 6-1. The complexity estimates resulted from assuming the use of commercial chips only. There are two FFTs. Each FFT requires six computational elements and five delay commutators for a total of 22 circuit cards. The inverse FFTs use the same complement of cards as the FFTs. Thus the FFT is the major contributor to the system complexity (44 of 49 circuit cards). Each circuit card is approximately 15 inches square. Since the delay commutator is the major contributor to the complexity with nearly 60% of the system chip count, attention has been focused on reducing its complexity.

6.4 Semi-Custom VLSI Delay Commutator Implementation

The high complexity of the delay commutator is a direct result of the difficulty of realizing shift registers that can be set to a variety of lengths as required for the various delays. The most efficient approach involves simulating a delay line by using a RAM with write and read addresses displaced by a constant (i.e., the length of the simulated delay line) as shown in Figure 6-15. In view of the high

Module Type	Modules	Chips
Data Acquisition and Weighting	2	240
FFT		
Computational Element	12	960
Delay/Commutator	10	1800
Frequency Domain Filter	1	140
Power Spectral Density	1	78
Combiner	1	143
Inverse FFT (same as FFT)	22	2760
Total	49	6121

Table 6-1. Adaptive Filter Complexity

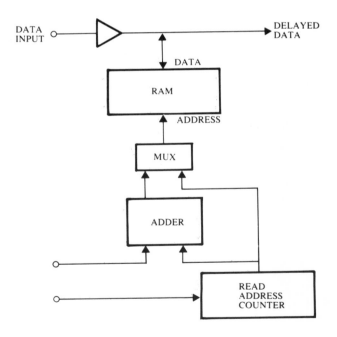

Figure 6-15. Delay Implementation

Frequency Domain Filter Case Study

complexity of the delay commutator, several VLSI implementation approaches (i.e., gate array, standard cell, and full custom) were considered. To effectively utilize technologies of widely varying logic density (i.e., gates/chip) a bit slice approach was selected for the 44 bit wide delay commutator. The complexity of an n bit wide slice is $400(n + 1)$ gates and $3,072\,n$ shift register stages. At an average equivalent complexity of three gates per shift register, the total equivalent complexity is approximately $10,000\,n$ gates. The options are compared in Table 6-2. Based on this comparison standard cell implementation was selected. A photograph of the resulting chip is shown in Figure 6-16. Since it is implemented as a four bit wide slice, the design consists of four nearly identical structures with an input shift register, commutation logic (realized with standard cells), and an output shift register. It achieves a 16 fold reduction in chip count at a three fold increase in development cost (relative to the commercial implementation). The further reduction in recurring complexity offered by custom design does not justify the greatly increased design cost. Similarly the gate array development cost savings are not great enough to justify the greater recurring complexity relative to the standard cell implementation. A goal is to coalesce the computational element and the delay commutator onto a single circuit board which is only feasible if the number of chips required to perform the delay commutator function is low relative to the 80 circuits for the computational element.

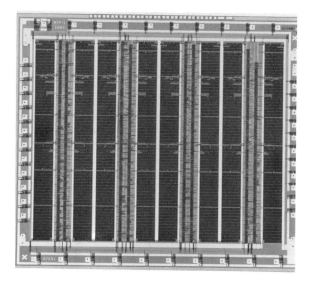

Figure 6-16. Delay Commutator Chip

	Gates/Chip	Max	Chips/Stage	Development Cost
Commercial	—	—	180	$25k
Gate Array	10K	1	44	$45K
Standard Cell	40K	4	11	$75K
Custom	100K	10	5	$300K

Table 6-2. Delay Commutator Implementation

By using standard cell semi-custom chips it is possible to eliminate the delay commutator module. The computational elements have enough room that the delay commutator chips can be added to the circuit board. The delay commutator chips are organized as 4 bit wide slices, where each processes 4 bits of the 44 bit data words (22 bit floating point complex pairs). As Table 6-3 demonstrates, use of the delay commutator chip reduces the FFT complexity by 60% [6-8].

A 4,096 point FFT requires the equivalent computation of 24,576 radix 2 butterfly computations (each consisting of four multiplication and six addition operations). Since this processor computes such transforms in 102.4 microseconds, it is computing 240 million butterflies per second or 2.4 billion arithmetic operations per second. Normalizing by the chip count of 546 gives a per chip functional throughput of 439,560 butterflies per second.

Since the 22-bit floating point adder consists of about 10,000 gates and the 22-bit floating point multiplier consists of about 6,000 gates, the radix 2 butterfly is equivalent to 84,000 gates. Multiplying by the butterfly rate gives the total equivalent gate-rate of 2×10^{13}. This is approximately 400 times the gate-rate of the commercial multiplier in Section 3.3, which is quite close to the actual chip count of 546. The average gate-rate per chip for the frequency domain filter is 3.7×10^{10}.

	FFT Complexity (Integrated Circuit Count)			
	Without Delay Communication Circuit		With Delay Commutator Circuit	
4096 Point FFT				
Computational Element	6 Cards at 80 CKTS/Card =	480 CKTS	6 Cards at 91 CKTS/Card =	546 CKTS
Delay Commutator	5 Cards at 180 CKTS/Card =	900 CKTS	(Included on CE)	
Total	11 Cards	1380 CKTS	6 Cards	546 CKTS

Table 6-3. Complexity Reduction Achieved with Delay Commutator Circuit

6.5 References

6-1 Fredric J. Harris, "On the Use of Windows for Harmonic Analysis with the Discrete Fourier Transform," **Proceedings of the IEEE,** Vol. 66, 1978, pp. 51-83.

6-2 Herbert L. Groginsky and George A. Works, "A Pipelined Fast Fourier Transform," **IEEE Transactions on Computers,** Vol. C-19, 1970, pp. 1015-1019.

6-3 Ben Gold and Theodore Bially, "Parallelism in Fast Fourier Transform Hardware," **IEEE Transactions on Audio and Electroacoustics,** Vol. AU-21, 1973, pp. 5-16.

6-4 Peter E. Blankenship and E. M. Hofstetter, "Digital Pulse Compression via Fast Convolution," **IEEE Transactions on Audio, Speech, and Signal Processing,** Vol. ASSP-23, 1975, pp. 189-201.

6-5 J.H. McClellan and R.J. Purdy, "Applications of Digital Signal Processing to Radar," in Alan V. Oppenheim, ed., **Applications of Digital Signal Processing,** Englewood Cliffs: Prentice-Hall, 1978, Ch. 5.

6-6 Earl E. Swartzlander, Jr. and John Eldon, "Arithmetic for High Speed FFT Implementation," **Proceedings Seventh Symposium on Computer Arithmetic,** Urbana, IL, 1985, pp. 223-230.

6-7 Lawrence R. Rabiner and Bernard Gold, **Theory and Application of Digital Signal Processing,** Englewood Cliffs: Prentice-Hall, 1975.

6-8 Earl E. Swartzlander, Jr., Wendell K.W. Young, and Saul J. Joseph, "A Radix 4 Delay Commutator for Fast Fourier Transform Processor Implementation," **IEEE Journal of Solid-State Circuits,** Vol. SC-19, 1984, pp. 702-709.

CHAPTER 7.
CUSTOM VLSI CASE STUDY

A detailed case study of custom VLSI development for an advanced digital beam forming system is presented in this chapter. The digital beam forming concept is explained in Section 7.1 to provide a basic understanding of the application.

Custom VLSI development involves merging activities in three areas: algorithms applicable to the application, architectures to implement the algorithms, and the capability of available technology. These activities interact, as shown in Figure 7-1, via design studies that culminate in an initial design and a feasibility assessment. Sections 7.2 through 7.4 examine algorithms, architecture, and the technology requirements to implement a digital beam former.

Given the initial design and an assessment of ultimate feasibility, use of a testbed as described in Section 7.5 facilitates the examination of many design and implementation issues (like arithmetic precision requirements, bus loading, etc.). The testbed is excited with simulated data to examine the functional performance under varying operating conditions. Design refinement studies conducted with the testbed produce a final VLSI design for implementation and specifications for the VLSI circuit. The VLSI payoff is summarized in Section 7.6.

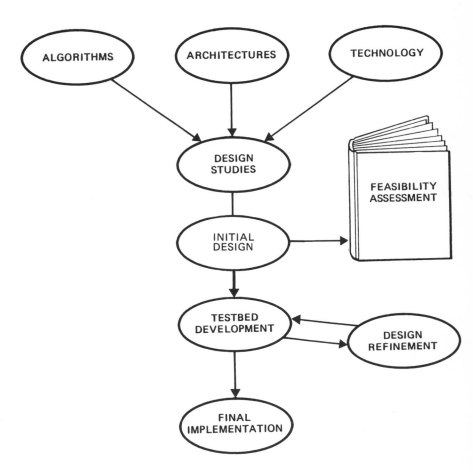

Figure 7-1. Custom VLSI Design Approach

7.1 Digital Beam Forming

Digital beam forming creates tailored antenna response patterns by receiving, sampling, digitizing, and combining data from many antenna elements. Based on the data, the arrival angles of the incoming signals are determined. Figure 7-2 shows a one-dimensional model of this process. For radar applications, a transmitter illuminates a volume of space, objects within the volume reflect energy, and a portion of the reflected energy impinges on the antenna array. The energy sensed by the antenna array is detected by receivers at each antenna element. The receiver outputs are digitized with A/D converters at each receiver. The digitized receiver outputs form a spatial sinusoidal pattern whose frequency depends on the angle of arrival of the radar return. The presence of an object at a specific direction is indicated by evaluating the corresponding component of the two dimensional spectrum.

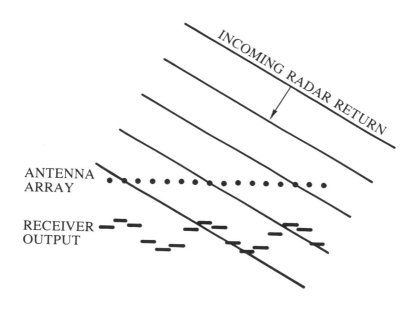

Figure 7-2. Beam Forming Concept

Thus the beam former operates as a spectrum analyzer to determine whether there are frequency components at specific frequencies corresponding to the angles of arrival (or beams) of interest. If only a single beam is to be formed, a single spectral component is examined. If multiple beams are of interest, multiple spatial frequencies are evaluated. To evaluate all possible beam patterns, a complete two-dimensional spatial spectrum of the digital data is formed. The choice between forming single beams, multiple beams, and all possible beams depends on the system application.

For radar applications, the beam former outputs are processed by radar signal processors. The radar signal processor outputs go to the user (e.g., an airport approach control system). Upon analysis of the beam output, the user provides control signals to the beam former giving the beam locations of interest for the next time interval.

The traditional phased array beam forming approach is shown in Figure 7-3. Radio frequency signals sensed by the antenna elements are coupled through phase shifters into a single receiver. The phase shifters are set to provide values of delay that electrically orient the antenna so that a maximum signal is obtained for a particular angular orientation [7-1]. The phased array system is attractive since it uses a single receiver and a single A/D converter; which is important when receivers and A/D converters are expensive relative to fabricating and calibrating the antenna and the phase shifter network. Each phase shifter needs to be located a specific electrical distance away from the antenna element it services. The distances have to be precisely matched, calibrated, and maintained, and all of the links from the phase shifters to the receiver have to be precisely matched.

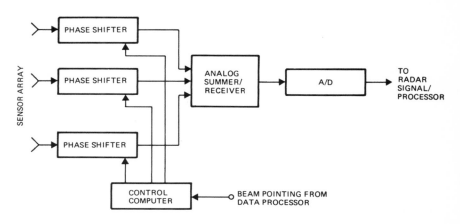

Figure 7-3. Phased Array Beam Forming

Because of the mechanical matching, phased arrays are more practical for single beams than for simultaneously forming several beams. At certain RF frequency bands the phase shifters exhibit slow settling times. If it is necessary to move a beam from one direction to another, a large amount of time may be required for beam reorientation.

For the digital approach, shown in Figure 7-4, a receiver and an A/D converter are placed at each antenna element. The resulting digital data are bused to the beam former which performs the spectral analysis. In contrast to the analog phased array approach, this scheme uses a receiver and an A/D converter for each antenna element. With the current state of technology, it is often easier and cheaper to replicate VLSI circuits than to construct precisely matched analog systems. The digital approach forms any number of beams simultaneously by simply evaluating multiple components of the spectrum. The simplicity of implementing data storage with low cost digital memory greatly facilitates the implementation of adaptive beam forming (see [7-2] through [7-4]) where observed data and noise are used to adapt the shape of the beam patterns.

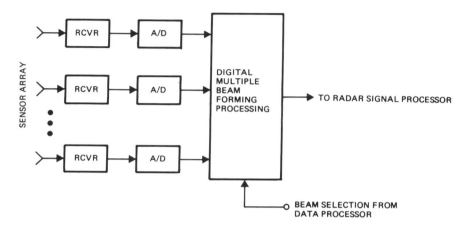

Figure 7-4. Digital Beam Forming

7.2 Algorithms

Algorithm selection is critical. Beam forming involves high speed two-dimensional spectrum analysis. The candidate algorithms include the Discrete Fourier Transform (DFT), the Fast Fourier Transform (FFT), and the Winograd Fourier Transform Algorithm (WFTA). Traditionally, the algorithms are compared on the basis of multiplication counts. As noted in Chapter 4, computing all frequency components (i.e., all possible beams) of an N sample data sequence with the DFT, requires N^2 complex multiplications. By use of the FFT algorithm, the complex multiplication count decreases to $(N/2)\log N$. Finally, with the WFTA it decreases to the order of N. Comparison on the basis of multiplication counts can be misleading for two reasons: it neglects other arithmetic operations and it assumes that control and memory complexity is negligible. This section focuses on total implementation complexity (not just the number of multiplies but the total computational load) which is a good indicator of the cost, size, power, reliability, etc.

A DFT structure to form one beam is shown in Figure 7-5. There is a complex multiplier for each digitized receiver output. The receiver outputs are multiplied by complex sine and cosine values to produce the desired phase shift and the products are summed to generate the beam output. This process is repeated for each beam, for each time sample. Thus, if the antenna array is M elements wide and N elements high there will be MN multiplications per beam. For K simultaneous beams the complex multiplication count is KMN per time sample. The number of complex additions is K(MN-1). There is negligible memory or control complexity. The only control requirement arises from the need to change the sine and cosine values whenever it is desired to change the direction of the beam that is being formed.

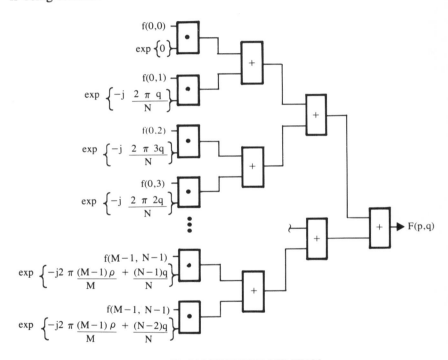

N · M MULTIPLIES PER BEAM

Figure 7-5. DFT Beam Former

Another approach is to use the FFT algorithm. The one-dimensional FFT is applied to the two-dimensional data array as shown in Figure 7-6 [7-5]. The N by M matrix of digitized receiver outputs is transformed on a column by column basis using M point FFTs to generate a matrix of intermediate results. Rows of intermediate results are transformed using N point FFTs to produce the beam patterns. If K beams are to be formed there need only be L row transforms where L is the number of rows of the output matrix that have at least one active beam, thus L = MIN(K,M). Each column transform requires an M point FFT which requires (M/2)logM butterfly computations. Since there are N columns the total is (NM/2)logM butterfly computations. The row transforms are N point FFTs for a total of (N/2)logN butterfly computations for each transform. With L row transforms, the total is (N/2)(MlogM + LlogN) butterfly computations. Since each butterfly computation requires one complex multiplication and two complex additions, the total arithmetic requirement is (N/2)(MlogM + LlogN) complex multiplications and N(MlogM + LlogN) complex additions. The FFT approach requires storage for the intermediate results so that a memory size of at least MN complex words is required. Control requirements include generation of the address sequences for the FFT algorithm and generation of the sine/cosine values.

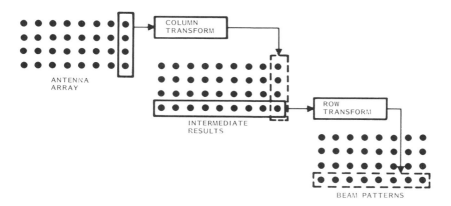

Figure 7-6. Two Dimensional FFT Beam Former

The WFTA approach is like the FFT approach in that the two-dimensional transform is performed via one-dimensional column and row transforms. The major difference is that N point one-dimensional transforms computed via the WFTA algorithm require approximately N complex multiplications and NlogN complex additions. Applying this to the approach shown for the FFT in Figure 7-6, the N column transforms involve NM complex multiplications and NMlogM complex additions. The L row transforms require LN complex multiplications and LNlogN complex additions. The total operation count is N(L + M) complex multiplications and N(MlogM + LlogN) complex additions. The WFTA memory and control requirements are much greater than for the FFT since the

148   VLSI Signal Processing Systems

WFTA is not an "in place" algorithm, meaning that data is accessed from memory, results are computed, but the results must be stored in locations other than where the data was accessed since the original data must be available for other computations.

The complexity of these three approaches to two-dimensional spectrum analysis are compared in Table 7-1. The table lists the number of arithmetic operations required to process one time sample from the M by N receiver array, indicates the amount of memory required by the three algorithms, and compares the relative control complexity. The DFT arithmetic requirements are proportional to the number of simultaneous beams, while the FFT and WFTA arithmetic requirements are only weakly dependent on the number of beams. As a result, the optimum algorithm (from arithmetic considerations) depends on the number of simultaneous beams. When a few beams are formed, the DFT requires the least arithmetic, while when many beams are formed, the FFT or WFTA requires the least arithmetic. The comparison of total complexity is more involved as it requires examination of the relative difficulty of arithmetic, memory, and control. Detailed design studies suggest that the DFT exhibits lowest total complexity when the number of simultaneous beams is less than 5 percent to 10 percent of the number of possible beams, and that the FFT complexity is lowest when more simultaneous beams are to be formed. In neither case is the WFTA optimum since its reduced multiplication (relative to the FFT) is not enough to compensate for its higher memory and control complexity. In data processing implementations with minicomputers where multiplication may be a very slow operation, the WFTA is often optimum, but such is not the case when special purpose hardware is being developed.

| | Arithmetic Operations | | Memory | Control |
	Complex Multiplitions	Complex Additions		
DFT	KMN	KMN-K	None	Low
FFT	$\frac{N}{2}(M \log_2 M + L \log_2 N)$	$N(M \log M + L \log N)$	MN	Moderate
WFT	$N(L + M)$	$N(M \log M + L \log N)$	$\gg MN$	High

Where   MXN = Array size
          K = Simultaneous beams
          L = Min (K, M)

Table 7-1. Implementation Complexity Comparison

Custom VLSI Case Study 149

7.3 Architecture

The DFT algorithm described in the previous section can be implemented directly with an assembly of complex multipliers and adders interconnected as shown on Figure 7-5. An M by N array of complex multipliers (supplied with appropriate sin/cos weights) is coupled into a tree of complex adders. Since each antenna element is to be sampled at a rate of S samples per second, the multiplier/adder structure must be capable of accepting data at an S sample per second rate. To form multiple beams, the multiplier/adder tree structure is either replicated with different sin/cos weights for each structure, or the speed is increased so that the structure processes data at an input rate of KS samples per second. In the latter case, a coefficient memory at each multiplier stores the sin/cos weights so that they can be cycled through the multiplier once for each input sample.

The first of these approaches implements a version of the architecture shown previously as the DFT spectrum analyzer in Section 4.3 (reproduced in Figure 7-7a). In this architecture the data from the M by N receiver array is sequenced onto a data bus that broadcasts the data to multiple DFT processors. Each DFT processor accepts the M by N data matrix, computes and sums products, and routes the beam response to an output processor. Depending on the relative speed of the DFT processors and the input data rate, multiple DFT processors are operated in parallel. As noted earlier in this section, if a single DFT processor is capable of forming a single beam during the time interval of a single set of input data, K processors can be operated in parallel to form K distinct beams.

A similar architecture can be used with FFT processors as shown in Figure 7-7b. Here it is assumed that the FFT processors are slower than the MNS input data rate so that multiple processors are necessary. The processors are operated in a time skewed mode as explained in Section 4.3.

7.4 Technology Requirements

The gate rate requirements can be calculated directly for the DFT beam former. For each sample, the DFT performs KMN complex multiplications and KMN-K complex additions. In estimating the gate rate a parallel structure with KMN multipliers and KMN-K adders is assumed. Each arithmetic element completes one calculation per sample. Since a complex multiplication is implemented with four real multiplications and two real additions and a complex addition is implemented with two real additions, 4KMN real multipliers and 4KMN − 2K real adders are required. As a first order estimate, realization of a j bit parallel fixed point multiplier requires approximately $20j^2$ two input gates, since the multiplier

150   VLSI Signal Processing Systems

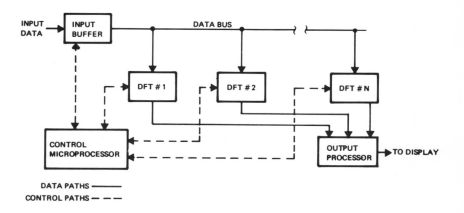

A. DFT Spectrum Analyzer

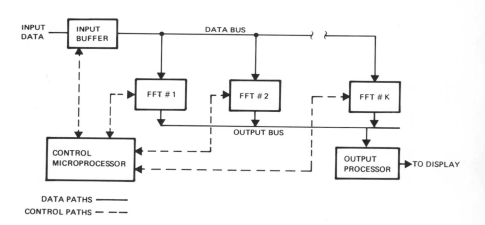

B. FFT Spectrum Analyzer

Figure 7-7. Spectrum Analyzer Architectures

Custom VLSI Case Study

requires j^2 full adders each consisting of 20 equivalent gates. Since a k bit ripple carry fixed point adder requires about 20k two input gates, the total gate requirement for the DFT beam former is $80KMN(j^2+k) - 40Kk$.

If all arithmetic elements have register for their inputs, the system clock rate is limited by the delay of the slowest arithmetic element. The required clock rate for the arithmetic elements is exactly equal to the sample rate of S samples per second. Thus, the necessary gate rate is $80\ KMNS(j^2+k) - 40KSk$.

The FFT beam former complexity is estimated by determining the number of gates to compute a two-dimensional transform. For the case where beams in all possible rows are of interest, $(MN/2)(\log MN)$ complex multiplications, $MN\log MN$ complex additions, and MN words of memory are required from Table 7-1. This converts into $2\ MN\log MN$ real multipliers of $20j^2$ gates, $3\ MN\log MN$ real adders of 20k gates, and MN complex words of memory of 8k gates for a total of $40\ MNj^2\log MN + 60\ MNk\log MN + 40\ kMN$ gates. To account for the control complexity of the FFT algorithm (address generation, sin/cos tables, etc.) this estimate of the complexity is doubled to $80\ MNj^2\log MN + 120\ MNk\log MN + 80\ kMN$ gates.

The required FFT speed of S transforms per second is achieved if the rate of the slowest element (i.e., multiplier, adder, or memory) is greater than S.

The FFT gate rate is the product of the FFT gate count times the required gate speed, which is $(80\ MNSj^2\log MN + 120\ MNSk\ \log MN + 80\ kMNS)$ where $k = 2j + 1 + \log MN$.

For the design feasibility assessment the estimates of the required gate rate to implement the DFT and FFT beam formers are compared with the capability of available technology to develop an estimate of the implementation complexity. This assessment requires estimation of the system parameters (i.e., antenna array size, A/D converter wordsize and sample rate, etc.). For one typical fixed radar application the parameters are: M = 16 elements, N = 32 elements, j = 10 bits, k = 30 bits, and S = 10^7 samples per second. For these values, the gate rates required for the DFT and FFT are 3.5×10^{13} and 5.5×10^{14}, respectively. Assuming the availability of technology that achieves a gate rate of 10^{12}, as shown in Figure 7-8, use of the gate rate metric suggests that a DFT beam former will require about 35 integrated circuits per beam. Similarly, a beam former developed using the FFT algorithm requires somewhat over 500 integrated circuits irrespective of the number of beams.

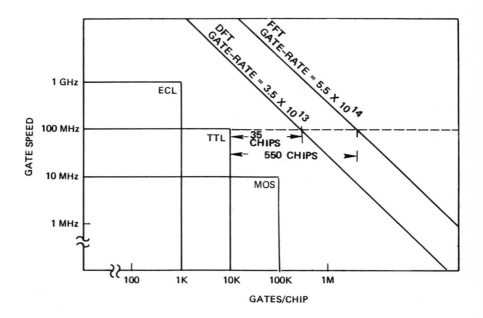

Figure 7-8. Digital Beam Forming Gate-Rate Requirements

After completing experiments with the testbed as described in Section 7.5, the DFT approach was selected for final implementation. The VLSI structure for the DFT beam former is shown in Figure 7-9. This is a complex multiplier accumulator which multiplies the digitized receiver outputs by sine/cosine values that are generated with a ROM. The ROM is addressed by a sequence generator (in this case, implemented with a simple accumulator) whose values govern which beam is formed. With four 10-bit multipliers, two 21-bit adders, and two 30-bit accumulators, the complexity is approximately 10,000 gates which is within the capability of the bipolar technology as shown in Figure 7-8. The DFT circuit is expected to operate at a 100 MHz clock rate. To process 512 samples at an average rate of 10 MHz approximately 50 DFT chips are required. The disparity between the actual requirement of 50 chips and the original estimate is well within the accuracy of the metric.

Custom VLSI Case Study

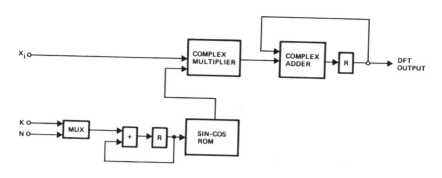

Figure 7-9. VLSI Discrete Fourier Transform Chip

7.5 Testbed

The goals of testbed development are to validate the architecture design and refine the technology selection [7-6]. In validating the architecture the issues include optimizing the data routing, tailoring the interconnection network to the processing elements, and distributing the computation. The testbed also simplifies network traffic loading estimation and protocol development.

It is also crucial to carefully refine the technology selection before beginning detailed VLSI circuit design. In circuit development, the chips must perform the right functions at the right speed with the right arithmetic precision. If 8 bit arithmetic is acceptable, but the system implements 16 bit arithmetic, the overall system will be far more complex (and far slower) than necessary. Similarly, if 12 bit arithmetic is necessary but 8 bit arithmetic is implemented the system won't work correctly. Either way, the mistake can be corrected easily prior to chip development. With the testbed, the chip design is refined by processing many different example situations to understand the round-off and overflow characteristics.

The testbed architecture is shown in Figure 7-10. A general purpose computer generates simulated radar data that is coupled through a controller onto a bus that distributes the data to each of the beam formers. The system can operate so that data is distributed in parallel to all the beam formers or is distributed to one processor at a time, depending on whether the DFT or FFT algorithm is selected. The results that are generated by the beam formers are coupled to the bus, through the controller, and to the computer for display.

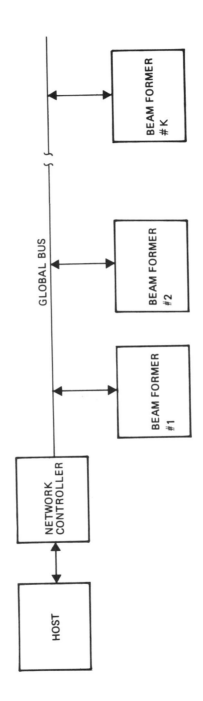

Figure 7-10. Digital Beam Forming Testbed Architecture

Custom VLSI Case Study

Three components are required to implement the testbed. The host computer generates simulated radar data, monitors and controls the experimentation, and provides data analysis and display of the beam former results. The network controller is an interface that couples the computer to the bus which is connected to the beam formers. Finally, the beam formers are single board microprocessors that provide the signal processing computation.

The beam former microprocessor design is shown in Figure 7-11. A local data bus connects the processor to program and data memory, a sine/cosine ROM, and a multiplier accumulator. The local bus is coupled through a bi-directional FIFO buffer to the global bus which connects to the other beam formers and to the network controller. Figure 7-12 is a photograph of the beam former.

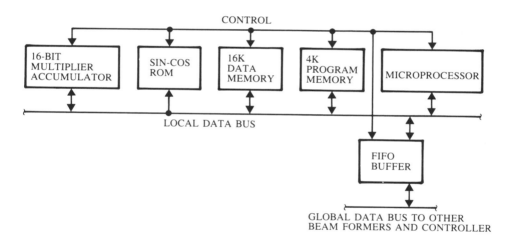

Figure 7-11. Testbed Beam Former Microprocessor

Figure 7-12. Testbed Beam Former Microprocessor

7.6 The VLSI Payoff

On the basis of system requirements, the initial design, and experimentation with the testbed, the DFT algorithm was selected for final implementation.

The most important issue in VLSI development is cost. If the VLSI based design costs less to implement than non-VLSI based designs then the benefits of the VLSI design will justify the required development effort. Figure 7-13 compares the development cost of the digital beam forming system against a phased array system with comparable performance. In 1978, the digital beam forming approach was about 25 percent more expensive. The development of single chip A/D converters caused the tradeoff to reverse by 1980. Improvements in digital technology (in part due to the influence of the VHSIC program) has caused the digital beam forming approach costs to continue to decline by a factor of four in the 1980 to 1985 period, while phased array costs remain relatively stable. This chart reflects the recurring production cost for a specific system that forms a small number of beams in parallel. Increasing the number of beams would improve the relative cost advantage of the digital beam forming approach since much of the cost is associated with the receiver and A/D converter array which doesn't change as additional beams are formed.

A secondary, but still important, advantage of the digital approach is the applicability of adaptive signal processing. This is a direct result of the ease of digital storage which allows the information contained in multiple returns to be exploited to provide greatly increased resolution, for example, by synthetic-aperture radar algorithms [7-7].

Custom VLSI Case Study

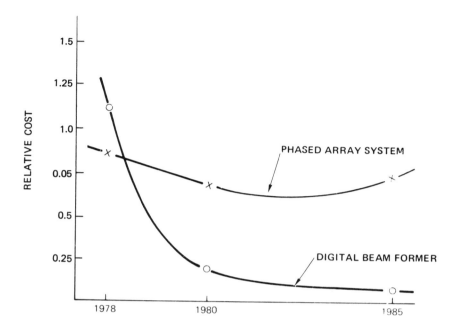

Figure 7-13. Beam Forming Cost Comparison

7.7 References

7-1 Robert A. Monzingo and Thomas W. Miller, **Introduction to Adaptive Arrays**, New York: John Wiley & Sons, 1980, Ch. 2, pp. 2-77.

7-2 C.F.N. Cowan and P.M. Grant, **Adaptive Filters**, Englewood Cliffs: Prentice-Hall, 1985.

7-3 Michael L. Honig and David G. Messerschmitt, **Adaptive Filters: Structures, Algorithms, and Applications**, Boston: Kluwer Academic Publishers, 1984.

7-4 Bernard Widrow and Samuel D. Stearns, **Adaptive Signal Processing**, Englewood Cliffs: Prentice-Hall, 1985.

7-5 Earl E. Swartzlander, Jr., "Signal Processor Design for Digital Beam Forming," **Proceedings of EASCON**, 1980, pp. 234-238.

7-6   Earl E. Swartzlander, Jr. and Joe M. McKay, "A Digital Beam Forming Processor," **Proceedings of the SPIE Real-Time Signal Processing Conference**, Vol. 241, 1980, pp. 232-237.

7-7   Kiyo Tomiyasu, "Tutorial Review of Synthetic-Aperature Radar (SAR) with Applications to Imaging of the Ocean Surface," **Proceedings of the IEEE**, Vol. 66, 1978, pp. 563-583.

CHAPTER 8.
SIGNAL PROCESSING NETWORKS

This chapter examines methods for the development of signal processing networks. The goal is to construct systems that achieve higher levels of throughput than are available with a single processor. Several different network types are described in Section 8.1 to illustrate the range of network topologies available to the designer. Section 8.2 develops criteria for comparing the performance, cost, and quality of the example networks. Finally, Section 8.3 presents an example of network development for a specific application and shows an extension of the general concepts presented in Sections 8.1 and 8.2.

At the present time, networking applies primarily to the development of large systems, but as the capability of VLSI increases, networking considerations will be an important aspect of circuit development. Ultimately (perhaps with Wafer Scale Integration?) the circuits themselves will have grown to such levels of complexity that they will consist of processors connected via on-chip networks. Most of the research in distributed computing and networking has been in the context of general purpose data processing, but the concepts apply directly to signal processing systems.

In this chapter a variety of signal processors (not necessarily of the same type) will be interconnected to implement collectively a larger task than any one processor can perform. The network serves as a data connection structure used for communication between the various processors. By using a network, the total computational load can be distributed amongst several processors. With a proper network structure and control philosophy, it is possibile to achieve fault tolerance so that even if one part of the system fails, the remainder survives and continues processing. During such degraded operation the throughput decreases because computing resources are lost, but the system still operates. There are, in fact many applications where the throughput requirements could be achieved with a single high performance processor, but a network of slower processors is used to provide fault tolerance. In applications such as missile or aircraft control, etc. such fault tolerance is absolutely crucial.

8.1 Common Network Types

This section describes several types of networks and characterizes their performance and complexity. Additional network types are examined in [8-1]. Two general network structures are widely used: processor-to-processor networks and resource sharing networks. In the processor-to-processor network, processors communicate through the network to other processors. Thus all network ports have processors connected to them. With resource sharing networks, the processors communicate through the network to shared resources such as memories where data is deposited for another processor to access. Both schemes offer equivalent connectivity, generality, and flexibility, but due to the development context, some network types are typically used in one mode or the other.

Crossbar Network

The crossbar network concept was initially developed for telephone and telecommunications message switching. Although originally implemented with mechanical switches, electronic versions were used to implement early multiprocessing computers by connecting serveral mainframe computers. Figure 8-1 shows the crossbar network. Along one edge are N processors; each is connected to a bus with N taps. Each tap consists of a switch that can be closed to connect that processor to one of the shared resources arrayed along the second edge of the crossbar. Consider a typical data transfer. Processor one wants to send data to shared resource two. Processor one determines if resource two is in use. If not, processor one closes the switch connecting its bus with the resource two bus. This establishes a direct connection: data is exchanged between processor one and

Signal Processing Networks

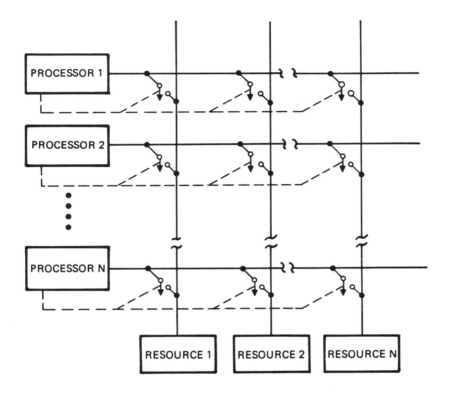

Figure 8-1. Crossbar Network

shared resource two as appropriate. When processor one is ready to access another shared resource, it goes through a similar process of interrogating the appropriate bus to determine if the resource is in use. If the resource is not in use, processor one can close the switch creating a direct connection. The crossbar network uses N^2 switches to implement the switching matrix for a system consisting of N processors and N shared resources. As a result, crossbars are frequently felt to be too expensive except for small networks.

Star Network

The star network shown in Figure 8-2 consists of a switching hub connected to a number of interface ports. The star switching hub connects data sources to destination ports. Although it is possible to use a crossbar to implement the network hub, most star networks are implemented with a single source selection switch that transfers data from any one source to any destination.

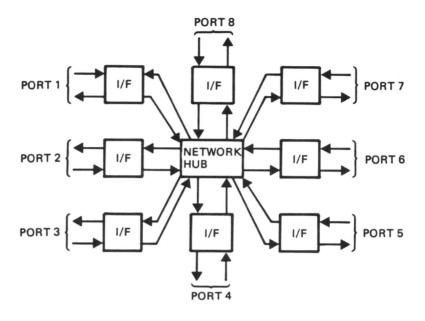

Figure 8-2. Star Network

The star network operation is shown in Figure 8-3. The input multiplexer selects a data source. That data source is connected through the network hub to a selector switch and to as many, or as few, of the destinations as appropriate. It is possible to implement common processing (e.g., signal regeneration, error detection, correction, etc.) in the hub [8-2]. At any time, one source is transmitting to any (or all) destination(s). If there is a failure within the central hub the entire network fails [8-3]. This single failure vulnerability is compensated by the relative simplicity of the star network which is implemented with 2N data links (one to and one from each node) and an N:1 multiplexer and a 1:N demultiplexer in the hub.

Cube Network

The cube network [8-4] and [8-5] is one of a family of related structures such as the perfect shuffle network [8-6], the shuffle exchange network [8-7], the reverse exchange network [8-8], the omega network [8-9], the delta network [8-10], the pi network [8-11], the Banyon network [8-12], and the Benes network [8-13], etc. The basic structures have been independently discovered by several researchers and shown to be equivalent in various studies [8-5], [8-11], and [8-14].

Signal Processing Networks

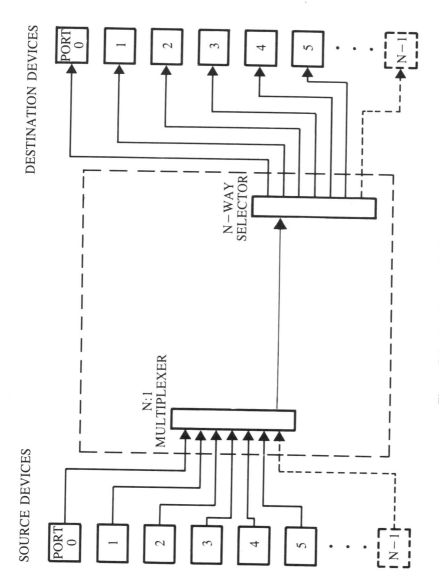

Figure 8-3. Star Network Hub Structure

The cube consists of log N columns of interchange boxes with N/2 boxes in each column as shown in Figure 8-4. At each interchange box the data passes straight through (i.e., connecting the upper input to the upper output and the lower input to the lower output) or is exchanged (i.e., the upper input is connected to the lower output and vice versa). For example, to send data from processor 7 to processor 5, the first column passes data straight through, the second column interchanges the data, and the third column passes data through. Any input can be routed to any output. The cube network is implemented with $N(1 + \log N)$ data links and $N/2 \log N$ exchange boxes.

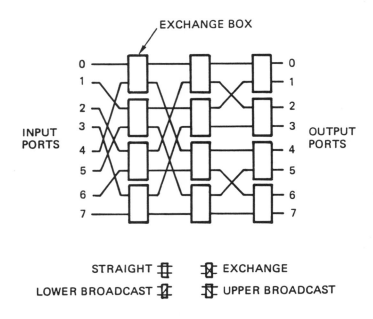

Figure 8-4. Cube Network

Ring Network

The ring structure involves connecting data links and interface nodes into a serial chain. The first interface node transmits to the second interface node, the second node transmits to the third node, and so on until the Nth node transmits to the first node [8-15], [8-16], and [8-17]. Ring structures, as shown in Figure 8-5, have the potential advantage that the data links that go from one processor to another are simple point-to-point links. A portion of a ring network shows

Signal Processing Networks

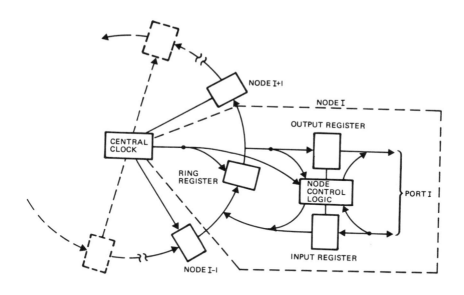

Figure 8-5. Ring Network

where there are four processors each connected to interface nodes. The interface nodes are shift registers that pass data from the Ith interface node to the I plus first interface node. Nodes are loaded with data from the processors, the data moves from node to node until it reaches the destination where it is unloaded. Ring structures are susceptible to failures in any interface node or data link. Since links are simple point-to-point links and since nodes are simple, high data transfer rates can be attained. The ring is implemented with N data links and N switching nodes (each consisting of a 1:2 multiplexer and a 2:1 demultiplexer).

One approach to circumvent the ring network susceptability to failure is to create a structure like the braided ring shown in Figure 8-6. At each interface node, a multiplexer is inserted at the node input. The multiplexer normally selects the output of the previous interface node, but can be switched to bypass the previous interface node (selecting in its place the second previous node). Such action eliminates the previous interface node from the ring. If that action solves the problem, the failure has been isolated and recovery procedures may be activated to restore correct system operation. If eliminating the previous node fails to solve the problem, then other nodes activate their multiplexers until the failure is eliminated. Relative to a simple ring network the braided ring adds N 2:1 multiplexers and doubles the number of data links to produce a structure that achieves a high degree of fault tolerance. The braided ring is implemented with

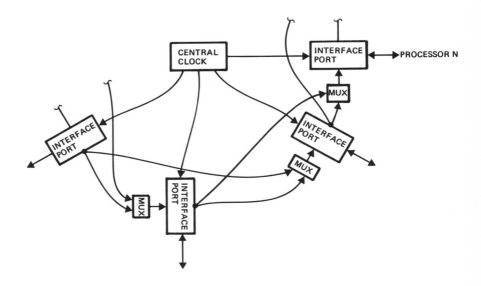

Figure 8-6. Fault Tolerant "Braided" Ring

N data links and 2:1 multiplexers in addition to the hardware required to implement a standard N node ring. Similar structures have been developed by adding "bypass" links to a standard ring, where the bypass links are chords that cut across the ring [8-18] and [8-19].

Bus Network

The bus structure shown in Figure 8-7 became popular because of its use on several early minicomputers to connect various peripherals to a single CPU [8-20]. In the general case a single data channel is shared by using switches to connect it to any of the processors. When a processor is ready to send a message, it tests to determine whether the bus is in use. If not, then the processor captures the bus and activates the appropriate switches sending messages to the selected destination. The bus is implemented with one data link and N SPST switches.

Fully Connected Network

The fully connected network uses direct point-to-point links from each processor to every other processor in the system [8-21]. This approach provides maximum throughput. The fully connected network consists of direct connections bet-

Signal Processing Networks

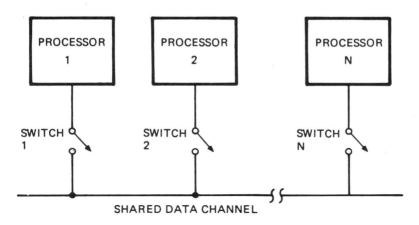

Figure 8-7. Bus Network

ween all processor pairs as shown in Figure 8-8. The number of links is given by L_{fc}:

$$L_{fc} = (N^2 - N)/2$$

Each node requires an N:1 multiplexer and a 1:N demultiplexer. As with the crossbar network, the complexity that grows as N^2 makes fully connected networks most popular for small systems.

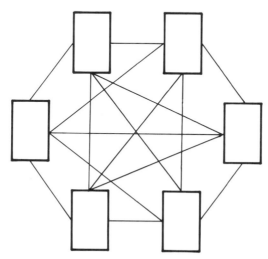

Figure 8-8. Fully Connected Network

Sparse Networks

Sparse networks like the example shown in Figure 8-9 are created by removing unnecessary data links from a fully connected network. Sparse networks are appropriate for special purpose systems where it is not necessary for every processor to access every other processor. A unique aspect of many special purpose networks is topological irregularity [8-22]. The system designer connects the network of processors in a flow form that is optimum for a specific problem, and can easily adjust the redundancy as appropriate for each specific data path through the network.

For special purpose processing, an irregular network that configures itself to match the algorithm flow is desirable. It is possible to distribute routing control amongst all switching nodes to eliminate the vulnerability of centralized control approaches and provide adaptive routing and graceful degradation. For comparison with other networks, it is assumed that there are an average of three data links per node in the sparse network and that the switch requirements are a 1:3 multiplexer and a 3:1 demultiplexer per node.

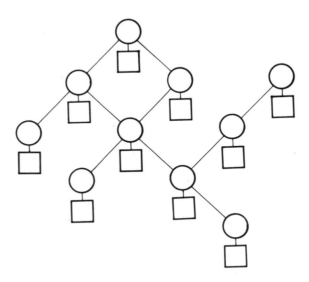

Figure 8-9. Example Sparse Network

8.2 Network Comparison

This section develops estimates for network performance and complexity. The network performance measure is defined as the number of simultaneous messages that can be processed by the network times the relative bandwidth of data links divided by the number of links that must be traversed to send a round trip message. The relation is:

NP = M B/D

Where: NP is the network performance, M is the number of simultaneous messages, B is the relative link bandwith, and D is the round trip message delay. Using relative link bandwidth reflects the disparity in communication rates between point-to-point links and multi-drop links. In the later case the bandwidth is degraded, partly due to the greater length and partly due to the physical taps which tend to cause increased line resistance and increased storage capacitance. A bus with K taps is assumed to operate at a bandwidth of 1/K relative to a point-to-point link. If 1/K is too severe a penalty, then other relations (for example $1/\sqrt{K}$, 1/log K, or even 1) can be easily incorporated into the comparison.

Table 8-1 presents the network performance for the networks described in Section 8.1. The crossbar, ring, fully connected, and sparse networks can all handle all N simultaneous messages. Cube networks can handle $\sqrt{N}$ simultaneous messages based on the probability of blockage [8-23]. For the bus and the star networks, there is only a single message at any time. The link bandwidth is one for all but the crossbar and bus networks where it is 1/N.

The round trip delay is the number of links that have to be traversed to go from processor A to processor B and back to processor A. In the crossbar network, data passes over two links when travelling from the source processor to the shared device. The two links are used to return data to the originating processor. Thus, the network performance is 1/4 (constant as a function of the network size for crossbar). Ignoring the bandwidth penalty would change the crossbar network performance to N/4 indicating growing capability as the network grows.

The star network has two link delays to go from one processor to another (i.e., one into the hub and another out to the destination) and another two delays for the return, for a total of four delays. Again, the network performance is constant at 1/4.

For the cube network, the number of stages to go from a processor to another processor is logN. Therefore, the total round trip delay is 2logN and the performance is $\sqrt{N}/(2 \log N)$ which is roughly constant for $2 \le N \le 128$.

Network Type	Number of Messages (M)	Link Band-Width (B)	Round Trip Delay (D)	Performance NP = MB/D
Crossbar	N	$1/N$	4	$1/4$
Star	1	1	4	$1/4$
Cube	$\sqrt{N}$	1	$2 \log N$	$\sqrt{N}/2 \log N$
Ring	N	1	N	1
Bus	1	$1/N$	2	$1/2N$
Fully Connected	N	1	2	$N/2$
Sparse	N	1	$\sqrt{N}$	$\sqrt{N}$

Table 8-1. Network Performance Results

For the ring, the delay from a processor to any other processor varies depending on where the processors are but the round trip delay is constant. This leads to a constant network performance of one.

The bus has the capability for only a single message, a bandwidth that decreases with increasing network size, and a constant round trip delay. Thus the network performance decreases at $1/2N$. Eliminating the penalty for multitap links would give a constant performance of $1/2$.

For the fully connected network the capability is N simultaneous messages, bandwidth of unity, and a round trip delay of two (one from processor A to processor B and another coming back) which gives network performance that increases as $N/2$.

For the sparse network, there can be N simultaneous messages, the link bandwidth is unity, and the average round trip delay is estimated at $\sqrt{N}$. Tailoring of the network connectivity to actual message routing should produce a lower average path length, but $\sqrt{N}$ is a conservative estimate. These characteristics give an estimated network performance of $\sqrt{N}$ indicating improving communication capability as the network size increases.

The network performance measure provides at least a crude estimate of the different networks' communication capabilities. In implementing a network based system, it is important to realize that network performance need only be as good as the system requirements. If the data rate is low or if only one processor ever transmits data at any time then a star network or a bus may be the best choice. Also, it has been assumed that the data access pattern avoids resource contention. If several processors try to access a single shared resource, the achievable network performance will be less than the estimated capability.

Cost

Two elements that comprise a network are data links and switches and accordingly cost can be measured by the number of data links and switches. For most purposes, links and switches are equivalent in cost. Switch complexity is reflected by taking the product of the number of switches, I, times the number of parallel lines through the switch, P, times the number of switch positions, T. The network cost, NC, is the sum of the number of links, L, and the switch complexity:

$$NC = L + IPT$$

In some cases this measure is not appropriate, For VLSI chip development, switches are relatively simple to implement and occupy very little area whereas the data links may require more area. In such cases it may be desirable to disregard the second term, i.e., IPT. In other cases it may be appropriate to ignore the data link cost. In either case weighting the two terms can be used to tailor the metric to the specific application.

Table 8-2 shows the network cost results for the example networks. The crossbar and the fully connected network have complexity that increases as N^2. In the crossbar this results from the switch complexity. For the fully connected network it results from both the increasing number of data links and the increased switch complexity. For the star, the ring, the bus, and the sparse networks the cost is roughly proportional to the network size. Finally, for the cube the cost grows in proportion to 3NlogN. Clearly, the crossbar and fully connected networks have costs that grow most rapidly. The cube, and the other network's cost grows at a rate proportionate to the number of network nodes.

Quality

Network quality, NQ, is defined as the network performance, NP, divided by the network cost, NC:

$$NQ = NP/NC.$$

Network Type	Links (L)	Switch Count (I)	Switch Type (P:T)	Switch Complexity S = IPT	Network Cost NC = L + S
Crossbar	2N	N^2	1:1	N^2	$N^2 + 2N$
Star	2N	1 N	N:1 1:1	2N	4N
Cube	$N + N \log N$	$(N \log N)/2$	2:2	$2N \log N$	$N + 3N \log N$
Ring	N	N	1:2 2:1	2N	3N
Bus	1	N	1:1	N	N + 1
Fully Connected	$(N^2 - N)/2$	N N	N:1 1:N	$2N^2$	$(5N^2 - N)/2$
Sparse	3N	N N	3:1 1:3	6N	9N

Table 8-2 Network Cost Results

The network quality results for the various network types are shown in Table 8-3. For the crossbar and the bus networks the network quality decreases in proportion to the size of the network as $1/N^2$. In both of these cases if the bandwidth penalty is ignored the degradation in network quality becomes proportional to $1/N$. For the star, the ring, and the fully connected networks, the network quality decreases in proportion to $1/N$. For the cube network the network quality decreases as $1/(\sqrt{N} \times \log^2 N)$. For the sparse network the network quality decreases in proportion to $1/\sqrt{N}$.

Figure 8-10 shows the network quality results over the range $N = 4$ to $N = 128$. These results confirm the intuitive notion that networks are less satisfactory as the number of processors increases. Networks are most effective when the number of processors is small. In general, the fully connected, sparse, and ring networks are better than the other networks for network quality, but the advantages are application dependent. For example, if there are many nodes in a network, the fully connected network represents overkill since each node has N-1 data links to other nodes but will only send one message at a time [8-3]. In many low throughput applications it is more appropriate to disregard the network quality comparison and instead minimize the network cost.

Signal Processing Networks

Network Type	Performance (NP)	Cost (NC)	Quality NQ = NP/NC
Crossbar	1/4	$N^2 + 2N$	$1/(4N^2 + 8N)$
Star	1/4	$4N$	$1/16N$
Cube	$\sqrt{N}/2 \log N$	$N + 3N \log N$	$1/(2\sqrt{N} \log N)(1 + 3 \log N)$
Ring	1	$3N$	$1/3N$
Bus	$1/2N$	$N + 1$	$1/(2N^2 + 2N)$
Full Connected	$N/2$	$(5N^2 - N)/2$	$1/(5N - 1)$
Sparse	$\sqrt{N}$	$9N$	$1/9\sqrt{N}$

Table 8-3. Network Quality Results

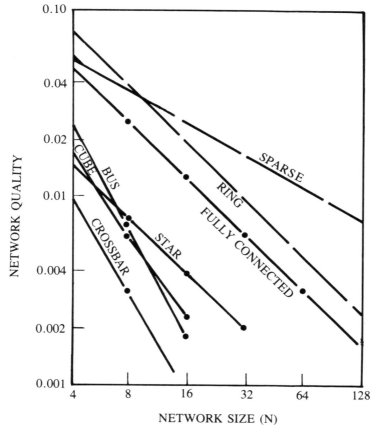

Figure 8-10. Network Quality Comparison

8.3 Network Design Example

With this introduction to network comparison it is appropriate to examine a specific example of network development. This example involves developing a large computing system. It will be implemented in several stages. Initially, a few processors will be connected where each processor simulates the eventual computing to be performed at that part of the system. Simulation will identify the processing bottlenecks. Then the processors that limit the system performance will be replaced with multiple processors to balance the loading.

The structure shown in Figure 8-11 was developed for this application. It is called the Gatlinburg Ring because the initial concept was first presented at an Army workshop in Gatlinburg, Tennessee in 1979 and amplified as discussed in [8-24] and [8-25]. The basic notion is to use two layers of rings. A high level ring is developed initially. At each high level ring node simulated data is processed with simulated timing programs to identify network bottlenecks. The most heavily loaded nodes are replaced by a second level ring which has additional computers that are interconnected to share the load of the high level ring node. Eventually all high level ring nodes will be replaced with second level rings, perhaps with variations in their size. This structure lends itself to evolutionary implementation where heavily loaded processors on the second level rings are replaced with third level rings of processors, etc. A similar concept was independently developed for national, regional, and local loops [8-15].

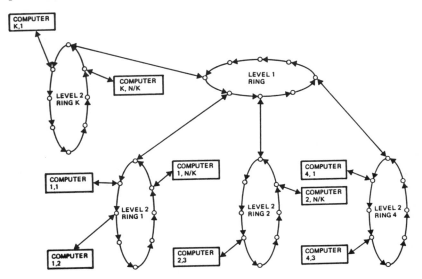

Figure 8-11. Gatlinburg Rings Network

To evaluate the network quality of the Gatlinburg Rings, it is necessary to select the number of low level rings. For simplicity, assume all low level rings are the same size and that there are K low level rings. This means there are N/K computers on each low level ring. The optimum value of K is that which produces the minimum round trip delay. The round trip delay, D_{RT}, is given by:

$$D_{RT} = K + 2N/K$$

Differentiation with respect to K and setting the derivitive to zero gives:

$$K = \sqrt{2N}$$

For this configuration the delay is $\sqrt{8N}$. Thus the network performance is:

$$NP = \sqrt{N/8}$$

The network cost is the same as for a conventional ring (3N). Combining these two gives the network quality, NQ:

$$NQ = \sqrt{72N}$$

Figure 8-12 compares the network quality of the Gatlinburg Rings with that of the conventional ring. Although not as good for small networks, the network quality of the Gatlinburg Rings degrades slower with increasing size and is better for large networks (i.e., when N is greater than 32). Other somewhat similar structures are clusters [8-26] and cones [8-27].

Networks provide the capability to adjust system performance by adding or removing computing elements without going through a complete system redesign. Such modularity means that systems can be modified easily to match changing system demands. A network quality estimate (i.e., performance/cost ratios) has been developed that agrees with empirical results. Specifically, network performance is adequate for small networks but decreases as the networks become larger.

There are several open issues in the networking area. First, there are few standard protocols, and those few are largely mutually incompatible. Second, in the development of distributed systems, data must be shared amongst the processors which requires procedures to ensure that processor A doesn't modify data that processor B is using, etc. Synchronization is also an issue. Even if all processors in a system are provided with a common clock signal, propagation delays will cause each processor to be operating slightly out of phase with respect to the others. This can cause problems in data intercommunication as well as control coordination. Each of these issues is important, but collectively they indicate the immature nature of the networking field. As solutions to these problems are developed and as technology attains even higher levels of integration, networking will become an issue of vital concern to VLSI designers.

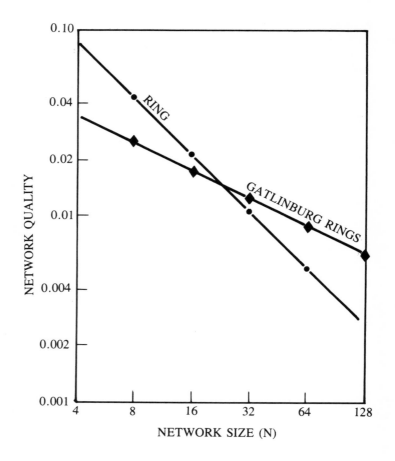

Figure 8-12. Gatlinburg Rings Quality Comparison

8.4 References

8-1   Tse-yun Feng, "A Survey of Interconnection Networks," **Computer**, December 1981, pp. 12-27.

8-2   Barry K. Gilbert, et al., "A Real-Time Hardware System for Digital Processing of Wide-Band Video Images," **IEEE Transactions on Computers**, Vol. C-25, 1976, pp. 1089-1100.

8-3 Larry D. Wittie, "Communication Structures for Large Networks of Microcomputers," **IEEE Transactions on Computers**, Vol. C-30, 1981, pp. 264-273.

8-4 Marshall C. Pease, III, "The Indirect Binary n-Cube Microprocessor Array," **IEEE Transactions on Computers**, Vol. C-26, 1977, pp. 458-473.

8-5 Howard J. Siegel, "The Theory Underlying the Partitioning of Permutation Networks," **IEEE Transactions on Computers**, Vol. C-29, 1980, pp. 791-801.

8-6 Harold S. Stone, "Parallel Processing with the Perfect Shuffle," **IEEE Transactions on Computers**, Vol. C-20, 1971, pp. 153-161.

8-7 Chuan-Lin Wu and Tse-yun Feng, "The Universality of the Shuffle-Exchange Network," **IEEE Transactions on Computers**, Vol. C-30, 1981, pp. 324-332.

8-8 Chuan-Lin Wu and Tse-yun Feng, "The Reverse-Exchange Interconnection Network," **IEEE Transactions on Computers**, Vol. C-29, 1980, pp. 801-811.

8-9 Duncan H. Lawrie, "Access and Alignment of Data in an Array Processor," **IEEE Transactions on Computers**, Vol. C-24, 1975, pp. 1145-1155.

8-10 Daniel M. Dias and J. Robert Jump, "Analysis and Simulation of Buffered Delta Networks," **IEEE Transactions on Computers**, Vol. C-30, 1981, pp. 273-282.

8-11 Pen-Chung Yew and Duncan H. Lawrie, "An Easily Controlled Network for Frequently Used Permutations," **IEEE Transactions on Computers**, Vol. C-30, 1981, pp. 296-298.

8-12 Mark A. Franklin, "VLSI Performance Comparison of Banyan and Crossbar Communications Networks," **IEEE Transactions on Computers**, Vol. C-30, 1981, pp. 283-291.

8-13 David Nassimi and Sartaj Sahni, "A Self-Routing Benes Network and Parallel Permutation Algorithms," **IEEE Transactions on Computers**, Vol. C-30, 1981, pp. 332-340.

8-14 Chuan-Lin Wu and Tse-yun Feng, "On a Class of Multistage Interconnection Networks," **IEEE Transactions on Computers**, Vol. C-29, 1980, pp. 694-702.

8-15 John R. Pierce, "How Far Can Data Loops Go?," **IEEE Transactions on Communications**, Vol. COM-20, 1972, pp. 527-530.

8-16 M.V. Wilkes, "Communication Using a Digital Ring," **Proceedings of the Pacific Area Computer Communication Network System Symposium**, Sendai, Japan, 1975, pp. 47-56.

8-17 Hossein Jafari, T.G. Lewis, and John D. Spragins, "Simulation of a Class of Ring-Structured Networks," **IEEE Transactions on Computers**, Vol. C-29, 1980, pp. 385-392.

8-18 Andrew Hopper and David J. Wheeler, "Binary Routing Networks," **IEEE Transactions on Computers**, Vol. C-28, 1979, pp. 699-703.

8-19 Bruce W. Arden and Hikyu Lee, "Analysis of Chordal Ring Network," **IEEE Transactions on Computers**, Vol. C-30, 1981, pp. 291-295.

8-20 John V. Levy, "Buses, The Skeleton of Computer Structures," in C. Gordon Bell, et al., eds., **Computer Engineering**, Bradford, MA: Digital Press, 1978, pp. 269-299.

8-21 David Katsuki, et al., "Pluribus – An Operational Fault-Tolerant Multiprocessor," **Proceedings of the IEEE**, Vol. 66, 1978, pp. 1146-1159.

8-22 Earl E. Swartzlander, Jr. and Douglas J. Heath, "A Routing Algorithm for Signal Processing Networks," **IEEE Transactions on Computers**, Vol. C-28, 1979, pp. 567-572.

8-23 Howard Jay Siegel, **Interconnection Networks for Large-Scale Parallel Processing**, Lexington, MA: Lexington Books, 1985.

8-24 Earl E. Swartzlander, Jr., "Distributed Signal Processing Systems," **Proceedings of the 14th Hawaii International Conference on Systems Sciences**, 1981, pp. 299-308.

8-25 Earl E. Swartzlander, Jr., "Networks for Embedded Computing," **Proceedings of the Third AIAA Computers in Aerospace Conference**, 1981, pp. 215-221.

8-26 Shyue B. Wu and Ming T. Liu, "A Cluster Structure as an Interconnection Network for Large Multimicrocomputer Systems," **IEEE Transactions on Computers**, Vol. C-30, 1981, pp. 254-264.

8-27 Leonard Uhr, **Algorithm-Structured Computer Arrays and Networks**, Orlando, FL: Academic Press, 1984, pp. 88, 94.

Author Index

E.J. Aas	10
Vishwani D. Agrawal	38
A.K. Agrawala	65
Sudhir R. Ahuja	116
Aristides G. Alexopoulos	116
Jonathan Allen	87
F. Anccau	10
Bruce W. Arden	178
D.F. Barbe	9, 10
Charles R. Baugh	116
C.G. Bell	64
Roderic Beresford	37
Frederic P. Beucler	38
Theodore Bially	139
P.E. Blankenship	116, 139
A.D. Booth	116
Willard Booth	38
B.A. Bowen	87
Don Brown	38
W.R. Brown	87
Randal Bryant	10
Arthur W. Burks	87
C. Sidney Burrus	86
Michael Burstein	37
Lynn Conway	65
James W. Cooley	86
C.F.N. Cowan	157
L. Dadda	115
S. Dasgupta	65
Norman G. Einspruch	10
John A. Eldon	38, 87, 139
Benjamin M. Elson	10
R. David Etchells	64
Tse-Yun Feng	176, 177, 178
Mark A. Franklin	177
S.H. Fuller	64
A. Ganesan	37
Gilles H. Garcia	116
Harvey L. Garner	116
Barry K. Gilbert	115, 176
Bernard Gold	38, 87, 139
Herman H. Goldstine	87
P.M. Grant	157
G. Grassl	38
John P. Gray	10
J.R. Grierson	37
Jan Grinberg	64
Herbert L. Groginsky	139
R.W. Hamming	86
Fredric J. Harris	139
Douglas J. Heath	178
Michael T. Heideman	86
George H. Heilmeier	64
P.J. Hicks	37
Eugene R. Hnatek	37
E.M. Hofstetter	139
Michael L. Honig	157
Andrew Hopper	178
M.J. Howes	10
De D. Hsu	38
K. Hwang	65
Hossein Jafari	178
Don H. Johnson	86
Saul J. Joseph	139
J. Robert Jump	116, 177
C.H. Kaman	64
David Katsuki	178

A.J. Kessler	37	Ralph Parris	38
N.G. Kingsbury	116	Ash M. Patel	37
Philip J. Klass	10	Marshall C. Pease, III	177
Robert K. Koin	37	Richard Pelavin	37
William J. Kubitz	116	A. Peled	87
H.T. Kung	10	John R. Pierce	178
		Robert L. Pritchard	36
Bernard S. Landman	37	R.J. Purdy	139
Duncan H. Lawrie	177		
Hikyu Lee	178	Lawrence R. Rabiner	38, 87, 139
Joseph Y. Lee	64	Charles M. Rader	87
V.R. Lessee	64	B. Randell	10
John V. Levy	178	T.G. Rauscher	65
T.G. Lewis	178	P.J.W. Rayner	116
Jeong-Tyng Li	37	Craig Robertson	87
B. Liu	87	Arthur L. Robinson	64
Ming T. Liu	179	Roy L. Russo	37
Michael J. Manner	38	Sartaj Sahni	177
Malgorzata Marek-Sadowska	37	Andres C. Salazar	87
J.H. McClellan	139	R.W. Schafer	87
E.J. McCluskey	38	Howard Jay Siegel	177, 178
Joe M. Mckay	158	Norman L. Soong	37
Carver Mead	65	John D. Spragins	178
David G. Messerschmitt	157	Bob Sproull	10
Thomas W. Miller	157	William W. Staley	38
Robert A. Monzingo	157	Samuel D. Stearns	157
E.F. Moore	64	Guy Steele	10
Gordon E. Moore	9, 64	William J. Stenzel	116
D.V. Morgan	10	Harold S. Stone	177
		Larry W. Sumney	9, 10, 36
David Nassimi	177	Earl E. Swartzlander, Jr.	36, 38, 64, 65, 87, 115, 116, 139, 157, 158, 178
John A. Nelson	38		
Dung Nguyen	37		
Robert N. Noyce	9	S.M. Sze	9
Graham R. Nudd	64		
H.J. Nussbaumer	87	Bou Nin Tien	37
		Benjamin S. Ting	37
A.V. Oppenheim	87	Kiyo Tomiyasu	158
		P.C. Treleaven	10
Kenneth P. Parker	38	Shuji Tsukiyama	37

Author Index

John W. Tukey	86
Leonard Uhr	179
Jeffrey D. Ullman	65
John von Neumann	87
Jerry Werner	10
David J. Wheeler	178
Bernard Widrow	157
M.V. Wilkes	64, 178
Thomas W. Williams	38
Beau R. Wilson	38
Larry D. Wittie	177
Bruce A. Wooley	116
George A. Works	139
Chuan-Lin Wu	177, 178
Shyue B. Wu	179
Pen-Chung Yew	177
Wendell K.W. Young	139

Subject Index

Accumulator 91, 113
Adaptive Beam Forming 145
Adaptive Data Routing 168
Adaptive Equalization 83-86
Adaptive Filtering 133, 135
Adaptive Signal Processing 156
Analog to Digital (A/D) Conversion
   49-51, 84, 125, 142, 144, 156
Architecture
  Arithmetic Function 47
  Harvard Computer 76-78
  Modular Signal Processor 78-82
  Programmable Signal Processor
  (PSP) 77-78
  Signal Processor 75-82
  VLSI 39-65
  von Neumann Computer 75-76
Arithmetic Instability 68
Array Multiplier Testing 31-36
Asynmetric Filter Kernel 91
Automated Routing 18

Banyan Network 162
Barker Code 68, 69
Beam Former Test Bed 153-156
Benes Network 162
Bit-Rate Metric 16
Bit Slice Microprocessor 82, 84
Booth's Multiplication Algorithm
   110-112
Braided Ring Network 165, 166
Bridging Circuit Faults 30
Bus Network 166-173

Carry-Lookahead Adder 96-99
Carry Save Adder 110, 111
Cell Library 20
Cellular Logic 6, 40

Cluster Network 175
Complementary MOS (CMOS)
  Technology 7
Computer Aided Design 16, 22, 39
Cone Network 175
Control Microprocessor 78-80,
   84, -86
Cooling 25
Correlation 67-69
Crossbar Network 160, 161,
   169-173
Cube Network 162, 164, 169-173
Custom VLSI 17, 23-25, 137, 138
   141-157

Dadda's Multiplication Algorithm 95
Data Link Bandwidth 169, 170
Data Overflow/Underflow 104,
   122, 153
Data Rounding 153
Data Windowing 118, 121
Delay Commutator Circuit 129-131,
   135, 137, 138
Delay Line 129, 130, 135, 136
Delta Network 162
Design Cost 39
Design for Testability 29, 30
Digital Beam Forming 141-145
Digital Filter 67, 68, 89-139
Discrete Fourier Transform (DFT)
  16, 70-72, 79, 80, 145, 146, 148-153
Distributed Routing Control 168
Dual-In-Line Package (DIP) 28, 29
Dynamic RAM 4
Dynamic Range 73, 74, 102,
   108, 114

Elastic Buffer 53

Emitter Coupled Logic (ECL)   93,
    94, 110, 115, 123, 125, 133
Encoding   67
Equivalent Gate Count   12, 14, 18

Fast Fourier Transform (FFT)   15,
    16, 70-75, 80-85, 117-138, 145-153,
Fast Multiplication   95-96
Fault Tolerence   160
FFT Delay Commutator   129-131,
    135-138
Filter
  Adaptive   133, 135
  Asymmetric Kernel   91
  Digital   67, 68, 89-139
  Finite Impulse Response (FIR)
      14, 67, 89-115, 117
  Frequency Domain   117-138
  Input Driven   109
  Kernel   68, 89, 90, 112, 113,
      118, 132
  Low Pass Recursive   68
  Output Driven   93
  Recursive   67, 68
  Symmetric Kernel   90
Finite Impulse Response (FIR) Filter
    14, 67, 89-115, 117
First In First Out (FIFO) Buffer
    53, 155
Floating Point
  Addition   126, 128
  Arithmetic   73, 74, 122, 126-129
  Multiplication   128, 129
Fourier Transform   118
Frequency Domain Filter   117-138
Full Adder   96-98, 101, 106, 112
Fully Connected Network   166-173
Functional Breakage   115
Functional Performance Metric
    12, 14, 94
Functional Synergy   27

Functional Throughput   74, 138
Functional Throughput Rate (FTR)
  Metric   12, 14

Gallium Arsenide   13
Gate Array   16-20, 23-25, 137, 138
Gate-Rate Metric   12-16, 47, 74,
    75, 138, 149-152
Gate Speed   12
Gatlinburg Rings Network   174-176
Global Data Bus   154, 155
Gold Standard Test Circuit   31
Graceful Degradation   168

Harvard Computer Architecture 76-78

Ill Conditioned Logic   55
Implicit Sequencing   57
Inner Product   91, 96
Input Driven Filter   109
Instruction Mix   14
Interconnection Area   18
Inverse FFT   118, 120, 123, 133,
    134

Large Scale Integration (LSI)   2-6
Latency   108-109, 114
Leadless Chip Carrier (LCC)   29
Level Sensitive Scan Design (LSSD)
    30
Local Data Bus   155
Logical Critical Mass   123
Low Pass Recursive Filter   68

Medium Scale Integration (MSI)
    1-5, 20, 27
Memory Bottleneck   76
Merged Arithmetic 93, 95-102, 115
Metal Oxide Semiconductor (MOS)
  Technology   4, 123
Microprocessor   4, 5

Subject Index

Microprogrammed Control 54-63, 86
Modular Signal Processor Architecture
    78-82
Moore Machine   54-60, 63
Moore's Law   1, 2
Multiplexed Data Ports 41, 45, 46
Multiplier Accumulator 41, 43, 44,
    82, 94, 95, 102, 113, 115
Multiport Memory 51-54, 81, 82

Network
  Benyan   162
  Benes   162
  Blockage   169
  Braided Ring   165, 166
  Bus   166-173
  Cluster   175
  Cone   175
  Cost   171-175
  Crossbar   160, 161, 169-173
  Cube   162, 164, 169-173
  Delta   162
  Fully Connected   166-173
  Gatlinburg Rings   174-176
  Perfect Shuffle   162
  Performance   169-175
  Pi   162
  Processor-to-Processor   160
  Quality   171-175
  Resource Sharing   160
  Reverse Exchange   162
  Ring   164-166, 169-176
  Shuffle Exchange   162
  Signal Processing   159-162
  Sparse   168-173
  Star   161-163, 169-173
NMOS Technology   13

Omega Network   162
Output Driven Filter   93
Output Encoding   60

Oxide Isolated ECL   110

Packaging   11, 25-29
Parallel Interfaces   44-45
Parallelism   70, 123
Parallel Multiplier 40-43, 46-49, 84
Perfect Shuffle Network   162
Performance Metrics   11, 12
Phased Array Beam Forming   144
Phase Shifter   144
Pi Network   162
Pin Grid Array (PGA)   29
Pin Out Requirements   26, 27, 47, 77, 100
Pipelined Adder   112
Pipeline FFT   125-127, 134
Pipelining   110, 112, 114
Power Consumption   27, 28
Power Spectral Density   118, 119, 132
Processor-to-Processor Network 160
Production Testing   29-36
Programmable Logic Array (PLA)
    60, 61
Programmable Signal Processor (PSP)
    77-78
Pseudo Noise (PN) Sequence   69

Radix 2 FFT Butterfly 15, 71, 74, 126
Radix 4 FFT Butterfly   126
Random Access Memory (RAM) 4, 76, 77, 132
Random Logic   55
Read Only Memory (ROM)   60, 61, 76, 77, 96, 105-107, 155
Recursive Averaging   132
Recursive Filter   67, 68
Rent's Rule   26, 27
Resourse Sharing Networks   160
Reverse Exchange Network   162
Ring Network   164-168, 169-176

Rounding	153	Transistor-Transistor Logic (TTL) 4, 27, 93, 94, 123, 133	
Scan/Set Logic	30	Two Port Memory	53
Second Sourcing	5		
Semi-Custom Design	11, 16-25, 135, 137, 138	Very High Speed Integrated Circuit (VHSIC) Program	
Serial Protocals	4, 5, 46	1, 9, 54, 123, 156	
Shuffle Exchange Network	162	Very Large Scale Integration (VLSI)	
Sign/Logarithm Arithmetic	102-108, 114, 115	1-7, 11, 26, 27 VLSI	
Signal Averaging	67, 69, 70	Architecture	39-65
Signal Processing	67-86	Design	6, 7, 11-38
Signal Processing Network	159-176	Development	7, 8
Signal Processor Architecture	75-82	Processing	9
Single Port Interfaces	48	von Neumann Computer Architecture	
Small Scale Integration (SSI)	1-5, 20, 27	75-76	
Sparse Network	168-173	Wafer-Scale Integration	159
Standard Cell Logic	17, 20-25, 137, 138	Winograd Fourier Transform Algorithm (WFTA) 145, 147, 148	
Star Network	161-163, 169-173		
Stuck Fault Model	30		
Subroutining	55, 59, 60		
Symmetric Filter Kernel	90		

Testability 6
Testing 11, 29-36
Time Domain Processing 67
Time-to-Market 25
Topological Irregularity 168
Transform
   Discrete Fourier (DFT) 16, 70-72, 79, 80, 145, 146, 148-153
   Fast Fourier (FFT) 15, 16, 70-75, 80-85, 117-138, 145-153
   Fourier 118
   Inverse 118, 120, 123, 133, 134
   Pipeline 125-127, 134
   Processing 67, 70-72, 84
   Winograd Fourier 145, 147, 148